Advanced Oxidation Handbook

Advanced Oxidation Handbook

James Collins
Jim Bolton, PhD

First Edition

American Water Works Association

Copyright © 2016 American Water Works Association.
All rights reserved.
Printed in the United States of America.

Project Manager/Senior Technical Editor: Melissa Valentine
Production: Andy Peterson

No part of this publication may be reproduced or transmitted in any form or by any means, electronic or mechanical, including photocopy, recording, or any information or retrieval system, except in the form of brief excerpts or quotations for review purposes, without the written permission of the publisher.

Disclaimer
This book is provided for informational purposes only, with the understanding that the publisher and authors are not thereby engaged in rendering engineering or other professional services. The authors and publisher make no claim as to the accuracy of the book's contents, or their applicability to any particular circumstance. The authors and publisher accept no liability to any person for the information or advice provided in this book, or for loss or damages incurred by any person as a result of reliance on its contents. The reader is urged to consult with an appropriate licensed professional before taking any action or making any interpretation that is within the realm of a licensed professional practice.

Library of Congress Cataloging-in-Publication Data
Names: Bolton, James R., 1937- author. | Collins, James, 1981, author.
Title: Advanced oxidation handbook / by James R. Bolton, PhD., James Collins.
Description: Denver, CO : American Water Works Association, [2016]
Identifiers: LCCN 2015050519| ISBN 9781583219843 | ISBN 9781613002629 (eISBN)
Subjects: LCSH: Water--Purification--Oxidation. |
    Water--Purification--Ozonization.
Classification: LCC TD461 .B65 2016 | DDC 628.3/51--dc23 LC record available at http://lccn.loc.gov/2015050519

ISBN   978-1-58321-984-3
eISBN   978-1-61300-262-9

American Water Works Association

6666 West Quincy Avenue
Denver, CO 80235-3098
303.794.7711

Contents

List of Figures ix
List of Tables xi
Preface xiii
Acknowledgments xv

Chapter 1 **Introduction** 1
What Is Advanced Oxidation? 1
What Are Possible Applications of Advanced Oxidation? 2
History of Advanced Oxidation 3
Government Regulations 4
Advantages and Disadvantages of Advanced Oxidation 5
AOT Handbook Organization 7
References 7

Chapter 2 **Fundamentals of Ultraviolet Light and Photochemistry** 11
Light and Photons 12
UV Light Spectral Ranges 12
Emission, Transmission, and Absorption of Light 14
Excited States and the Absorption Process 17
Laws of Photochemistry 19
References 20

Chapter 3 **Fundamentals of Advanced Oxidation** 23
Properties of the Hydroxyl Radical 23
AOT Mechanism 24
AOT Experiments in a Collimated Beam 25
Sensitized Photochemical Reactions 26
Rates of Direct Photolysis Reactions With Monochromatic Light 29
AOT Experiments in a Stirred Tank or Merry-Go-Round Reactor 33
Summary Regarding Sensitized Photochemical Reactions 33
AOT Figures-of-Merit 34
AOT Examples 39
References 48

Chapter 4	**Advanced Oxidation Types** 51	
	Light-Driven Homogeneous AOTs 51	
	Direct Photolysis 51	
	Dark Homogeneous AOTs 55	
	Light-Driven Heterogeneous AOTs 56	
	Homogeneous Advanced Reduction Processes 57	
	Summary 57	
	References 59	
Chapter 5	**Advanced Oxidation Equipment** 61	
	Available UV Equipment 61	
	UV Lamps 62	
	UV Sensors 66	
	Sleeves 67	
	Other Components 67	
	Available Ozone Equipment 68	
	Ancillary AOT Equipment 71	
	References 72	
Chapter 6	**Effects of Water Quality on AOT Systems** 73	
	UV Transmittance and Absorbance 73	
	Hydroxyl Radical Scavenging Demand 76	
	Turbidity 77	
	Disinfectant Residual 77	
	UV Lamp Sleeve Fouling 78	
	By-broducts From AOT Treatment 79	
	References 82	
Chapter 7	**Potential Locations for AOT Facilities** 85	
	Surface Water Applications 85	
	Groundwater Applications 87	
	Reuse Applications 88	
	References 89	
Chapter 8	**AOT System Design Considerations** 91	
	Treatment Goals 91	
	Key Design Criteria 91	
	UV AOT Design Criteria 94	
	Ozone Key Design Criteria 97	
	Chemical Feed System Design 99	
	Hydrogen Peroxide Quenching 101	
	Hydraulics 104	
	Electric Power Systems 105	
	Site Constraints/Layout 106	

	Cost Estimation 108	
	Treatability Testing 109	
	References 110	
Chapter 9	**Start-up, Operations, and Maintenance**	**111**
	Steps for Facility Start-up 111	
	Routine Operations and Maintenance Tasks 115	
	Monitoring 119	
	Energy and Chemical Management 122	
	References 124	
Chapter 10	**AOT Case Studies**	**125**
	Treatment of Taste and Odor 125	
	Treatment of Micropollutants 126	
	Reuse Treatment 132	
Chapter 11	**Safety and Handling of AOT Equipment**	**137**
	Electrical Safety 137	
	UV Light Exposure 138	
	Burn Safety 138	
	Lamp Break Issues 138	
	Mercury Release Response 141	
	Chemical Safety 143	
	References 145	
Chapter 12	**Considerations for a Water Utility Manager**	**147**
	Using an Engineering Consulting Firm 147	
	Information Needs 147	
	Questions for a Potential Engineering Consulting Firm 148	
	Questions for a Potential Equipment Manufacturer 149	
	Government Regulations 151	
	References 151	
Appendix A	**Terms, Units, Symbols, and Definitions** 153	
Appendix B	**Rate Constants and Quantum Yields** 161	
Appendix C	**Calculation of Fraction of UV Absorbed for UV/H_2O_2 AOT** 167	

Index 171

About the Authors

Figures

2-1	Spectral ranges of UV, visible, and IR light.	13
2-2	Reflection and refraction as a light beam passes from medium 1 with refractive index n_1 to medium 2 with refractive index n_2.	15
2-3	(a) Discrete molecular absorption: molecular energy level diagram showing excitation from the ground S_0 singlet state to either the S_1 or S_2 excited singlet state; and (b) delocalized absorption (as in a semiconductor): electronic transition in a semiconductor from the lower valance band (vb) to the upper conduction band (cb).	18
3-1	(a) CB1 collimated beam apparatus and (b) CB2 commercial collimated beam apparatus	27
3-2	Masked collimated beam setup for photochemical studies	28
3-3	Photolysis decay of N-nitrosodimethylamine (NDMA) in aqueous solution	32
3-4	Plots of log(c) versus E_{EO}: (a) photolysis of NDMA and (b) UV/H_2O_2 treatment of 1,4-dioxane	35
3-5	Calculated E_{EO} for an annular UV reactor (one UV lamp) as a function of the radius of the reactor and the percent transmitted (%T) of the water	37
3-6	Log of the concentration of methylene blue versus EED	38
3-7	Photolysis decay of NDMA and the growth of several products.	39
3-8	E_{EO} as a function of the H_2O_2 concentration in the UV/H_2O_2 degradation of MTBE	41
3-9a	Concentration of MTBE and some of the intermediates as a function of time	42
3-9b	Concentration of MTBE and selected intermediates as a function of time	42
3-9c	Concentration of MTBE and some of the carboxylic acid intermediates as a function of time	43
3-10	Concentration of H_2O_2 and the calculated and measured TOC for the UV/H_2O_2 degradation of MTBE as a function of time	43
3-11	Comparison of the absorption spectra of ferrioxalate and hydrogen peroxide with the emission spectrum of a medium-pressure UV lamp	44
3-12	Comparison of the UV-vis/ferrioxalate/H_2O_2 process and the UV/H_2O_2 process for the treatment of a water contaminated with 15 mg/L of BTEX	45
3-13	Experimental setup for the treatment of MTBE by the O_3/H_2O_2 AOT	45

3-14	Treatment of MTBE-spiked tap water and an MTBE-contaminated groundwater at the bench scale with the O_3/H_2O_2 process	46
3-15	Apparatus for the treatment of an MTBE-contaminated groundwater at the pilot scale with the O_3 and O_3/H_2O_2 processes	47
3-16a	Percent removal of MTBE in an MTBE-contaminated groundwater using the O_3/H_2O_2 process and O_3 alone for an MTBE concentration of 2.23 mg/L	47
3-16b	Percent removal of MTBE in an MTBE-contaminated groundwater using the O_3/H_2O_2 process and O_3 alone for an MTBE concentration of 0.18 mg/L	48
4-1	Absorption spectra of some pollutants [trichloroethylene (TCE) and N-nitrosodimethylamine (NDMA)] and the emission spectrum of a medium-pressure UV lamp	52
5-1	Closed-pipe UV reactor types: (a) multiple lamp reactor with lamps parallel to flow and (b) multiple lamp reactor with lamps perpendicular to flow	63
5-2	Relative spectral emittance from low-pressure and medium-pressure lamps	65
8-1	Example of an ozone demand and decay curve at 25°C	98
8-2	Total trihalomethane formation with sodium hypochlorite and sodium sulfite peroxide quenching	102
10-1	Sung-Nam Water Treatment Plant ozone generator and hydrogen peroxide storage system	127
10-2	Si-Heung Water Treatment Plant UV reactor and hydrogen peroxide storage system	128
10-3	Si-Heung Water Treatment Plant UV/H_2O_2 full-scale performance test results	128
10-4	Metaldehyde reduction as a function of ozone dose	129
10-5	Illustration of potential full-scale ozone/peroxide system	129
10-6	UV reactors and GAC contactors used at the TARP AOP water treatment facility	130
10-7	1,4-dioxane removal through UV/H_2O_2 treatment	132
10-8	TCE removal through UV/H_2O_2 and GAC	132
10-9	Hydrogen peroxide removal through UV/H_2O_2 and GAC	133
10-10	CFD-I model used to predict performance at the City of Scottsdale AWT and the UV reactor installed	134
10-11	City of Scottsdale AWT performance test results	135
A-1	Illustration of the concepts of irradiance and fluence rate: (a) irradiance onto a surface; (b) fluence rate through an infinitesimally small sphere of cross-sectional area dA	156

Tables

1-1	Advanced oxidation history	3
1-2	Example US state guidelines for 1,4-dioxane (as of June 2015)	5
1-3	USEPA suggested guidelines for water reuse	6
2-1	Spectral ranges of interest in photochemistry	13
3-1	Typical electrical energy per order values for some groups of pollutants	36
4-1	Comparison of AOTs	58
5-1	Classification of UV lamps	63
5-2	Emission wavelengths for some common excilamps	65
5-3	Comparison of the characteristics of UV lamps used for UV disinfection of drinking water	66
5-4	Ozone instrumentation	71
6-1	Summary of the molar absorption coeffcients at 254 nm for components that may be present in drinking water	75
7-1	Filtered surface water AOT locations	87
8-1	Ozone injection advantages and disadvantages	99
8-2	Comparison of sodium hypochlorite and GAC for peroxide quenching	103
9-1	Example operations and maintenance tasks	116
9-2	Example monitoring frequencies for key UV AOT operational parameters	122
9-3	Example monitoring frequencies for key ozone AOT operational parameters	123
10-1	Sung-Nam Water Treatment Plant ozone/H_2O_2 design criteria	127
10-2	Sung-Nam Water Treatment Plant UV/H_2O_2 design criteria	128
10-3	TARP UV/peroxide and GAC design criteria	131
10-4	City of Scottsdale AWT UV photolysis design criteria	133

11-1	Summary of online lamp break causes and prevention methods	140
11-2	Health and safety standards for mercury compounds in air	141
11-3	Mercury sampling locations	144
11-4	Health and safety standards for ozone in air	145
12-1	Possible content for AOT equipment specifications	149
A-1	Physical constants of interest in ultraviolet technologies	153
B-1	·OH radical rate constants	161
B-2	Quantum yields	163
C-1	Calculations for the fraction of UV absorbed by H_2O_2	167

Preface

Advanced oxidation technologies (AOTs) involve the use of powerful oxidizing intermediates (e.g., the hydroxyl radical ·OH) that can oxidize and degrade primarily organic pollutants from contaminated air and water. The term *advanced* is used because the chemical reactions involved are essentially the same (except billions of times faster) as the reactions that would occur if these pollutants were exposed in a natural environment. AOTs oxidize a broad range of contaminants, including those that are not readily removed with other advanced technologies. Interest in AOTs has grown in recent years with the need to treat an ever expanding range of regulated and emerging contaminants.

This handbook provides a brief introduction (with examples) to the concepts related to the fundamentals, design, and operation of AOTs. It is designed to help the beginner and to provide important reference material for those experienced with AOTs; however, this handbook is not intended to be an exhaustive review of all concepts related to AOTs.

No background is assumed except that of general science, such as that given in undergraduate science and engineering degree programs. This handbook should be of value to engineering and scientific consultants, water treatment operators and managers, students and faculty members in science and engineering programs that deal with water treatment, and staff in government regulatory offices.

Careful attention has been paid to terminology, units, and definitions. Appendix A provides a convenient summary.

Acknowledgments

The authors of this book gratefully acknowledge Arcadis U.S. Inc for supporting the time to write this book for James Collins and Chris Poepping (Arcadis employees). Chris Poepping helped write multiple chapters of the book, and without him, this book would not have been possible.

The writing of this book also required the help of many friends and colleagues who read and commented on various chapters. Christine Cotton, Caroline Russell, Corin Marron, Dan Olsen, and Bree Carrico provided important support and feedback as internal reviewers. We are also particularly grateful to Adam Festger of Trojan Technologies and Keith Bircher of Calgon Carbon Corporation, who acted as external reviewers for the book.

The authors would also like to thank the key contributions from WEDECO, a Xylem Brand; Trojan Technologies; Calgon Carbon Corporation; K-Water; Anglian Water; Tucson Water; and City of Scottsdale for providing important case study information.

James Collins would like to thank his wife, Marcia Moreno Baez, for supporting him through the duration of developing this book. Jim Bolton would like to thank his wife, Ingrid Crowther, for her unwavering support and encouragement during the writing of this book.

1

Introduction

WHAT IS ADVANCED OXIDATION?

Advanced oxidation technologies (AOTs) involve the use of powerful oxidizing intermediates (e.g., the hydroxyl radical ·OH) that can oxidize and degrade primarily organic pollutants from contaminated air and water. The term *advanced* is used because the chemical reactions involved are essentially the same (except billions of times faster) as the reactions that would occur if these pollutants were exposed in a natural environment. AOTs oxidize a broad range of contaminants, including those that are not readily removed with other advanced technologies (e.g., reverse osmosis or granular activated carbon).

Most of the commercially viable AOTs use either ozone or photochemical processes [i.e., ultraviolet (UV) or visible light] to generate ·OH radicals. Although conventional ozone treatment relies on oxidation, ozone treatment alone is not considered an AOT. Ozone-based AOTs would include ozone combined with hydrogen peroxide or UV to form hydroxyl radicals. This handbook will present a range of AOTs, but the focus is limited to UV and ozone-based AOTs as they are the most commonly used AOTs in municipal treatment applications.

Treatment with AOTs leads not only to the destruction of the target pollutants susceptible to oxidation but also, given sufficient treatment time, to complete mineralization (i.e., the only products are CO_2, H_2O and mineral acids [e.g., HCl, HNO_3, H_2SO_4, etc.] for any Cl, N, S, etc. present in the pollutants) of the pollutants and their by-products. However, because the intermediate products of AOT reactions are often nontoxic and/or readily biodegradable with biological treatment (Linden et al. 2015), treatment to complete mineralization, in most cases, is not necessary nor cost-effective.

This handbook provides a brief introduction (with examples) to the concepts related to the fundamentals, design, and operation of AOTs. It is designed to help the beginner and to provide important reference material for those experienced with AOTs; however, this handbook is not intended to be an exhaustive review of all concepts related to AOTs. The following are recommended references for additional information:

- AOT reviews
  - Bolton and Cater (1994)
  - Legrini et al. (1993)
  - Ikehata and Gamal El-Din (2006)
- AOT books
  - Braun et al. (1991)
  - Oppenländer (2003)
  - Tarr (2003)
  - Parsons (2004)
- UV light fundamentals and photochemistry
  - Wayne and Wayne (1996)
  - Bolton and Cotton (2008)
  - Bolton (2010)

WHAT ARE POSSIBLE APPLICATIONS OF ADVANCED OXIDATION?

AOTs may be considered for treatment of many source waters to oxidize contaminants. However, there are three principal applications where AOTs provide effective treatment and are cost-effective when compared to other treatment technologies (e.g., granular activated carbon, membranes, etc).

Micropollutant Treatment

Micropollutants are pollutants present in water at microgram-per-liter (µg/L) or lower concentrations. These include volatile organic compounds (VOCs), pesticides, herbicides, endocrine-disrupting compounds, personal care products, pharmaceuticals, and so on. Most micropollutants are not easily treated in conventional water treatment processes, and there is concern about their potential health effects. For example, endocrine-disrupting compounds, at certain doses, are known to interfere with the hormone system of mammals and can cause cancerous tumors, birth defects, or developmental disorders. The solvent stabilizer 1,4-dioxane is an example of a micropollutant that is not readily removed or oxidized with other treatment technologies.

Treatment of Taste-and-Odor Compounds

During warmer months, some drinking water sources are subject to algal blooms that generate taste-and-odor compounds, such as geosmin and 2-methylisoborneol (MIB). Although these compounds do not represent a health hazard, they are detectable to customers at concentrations in the nanogram-per-liter range (ng/L) because they give drinking water an unpleasant taste and odor. Algal blooms may also be accompanied by the presence of algal toxins (e.g., microcystin) that do have known health effects.

Table 1-1 Advanced oxidation history

Date	Development	Reference
1894	Description of decomposition of tartaric acid by the addition of hydrogen peroxide (H_2O_2) to solutions containing ferrous ion (Fe^{2+}) at a pH of about 3. This is now known as the *Fenton reaction*.	Fenton 1894
1900	H_2O_2 observed to be decomposed by light.	Kistiakowsky 1900
1929	Proposed that the UV photolysis of H_2O_2 yields ·OH radicals.	Urey et al. 1929
1956	Described the photolysis of ozone in solution, determined the quantum yield, and identified O_2 and H_2O_2 as the products.	Taube 1956
1957	Determination that the quantum yield of H_2O_2 photolysis is 1.0.	Baxendale and Wilson 1957
1968	Description of products of the vacuum UV (VUV) photolysis of water.	Getoff and Schenck 1968
1975	Proposal of a mechanism for the decay of ozone that involved a pathway for the generation of ·OH radicals.	Hoigné and Bader 1975
1979	Introduction of the O_3/H_2O_2 process.	Nakayama et al. 1979
1982	Description of the UV/O_3 oxidation of trichloroethylene	Peyton et al. 1982
1987	First comprehensive reviews of advanced oxidation processes involving ozone, hydrogen peroxide, and ultraviolet light. Introduction of the term *advanced oxidation technologies* (AOTs).	Glaze et al. 1987
1996, 2001	Introduction of the energy efficiency concepts of electrical energy per order and electrical energy per mass.	Bolton et al. 1996, 2001
2012	Introduction of UV dose scale-up approach for UV-based AOTs.	Bircher et al. 2012

Recycled Water Treatment

There is increasing interest in the reuse and recycling of wastewater. Most wastewater contaminants can be removed from secondary effluents using membranes; however, many micropollutants cannot be completely removed by such treatments [e.g., 1,4-dioxane or *N*-nitrosodimethylamine (NDMA)]. Recycled water applications can also have a range of contaminants that are not typically found in most drinking water applications and AOTs can provide a broad treatment barrier for these contaminants, especially if the water is indirectly or directly augmenting potable water supplies.

HISTORY OF ADVANCED OXIDATION

The book by Oppenländer (2003) has a compilation of historical events related to the development of AOTs. Table 1-1 presents some important landmarks related to AOTs.

GOVERNMENT REGULATIONS

Ozone- and UV-based AOTs involve similar technologies and equipment as used for disinfection applications. Disinfection applications have well-defined standards and regulations that can be used to dictate equipment sizing. These standards include ozone concentration-detention time requirements (i.e., CT tables) and UV dose tables. Standard guidance for the design, operation, monitoring, and reporting of ozone and UV disinfection facilities can be found in the US Environmental Protection Agency (USEPA) guidance manuals. However, there are no USEPA guidance manuals available to aid in the design and operation of AOT facilities. The design of AOT systems are site-specific, dependent on the target contaminants and water quality, and vary by process and system manufacturer.

Water treatment at many utilities may be impacted by regulated or unregulated contaminants that can be effectively treated by AOTs. As part of the Safe Drinking Water Act, the USEPA has established maximum contaminant levels (MCLs) for a wide range of contaminants. Some of these contaminants are difficult to treat with conventional methods but can be effectively treated with AOTs (e.g., atrazine). For AOT applications targeting regulated contaminants, compliance is based on maintaining concentrations below the MCL based on the defined monitoring frequency. Therefore, the utility has flexibility in the design and operation of the AOT facility as long as concentrations are maintained below the MCL. For AOTs targeting unregulated contaminants (e.g., taste and odor), there is no USEPA required monitoring or reporting, and the system design and operation would then be based on utility preferences and any requirements of the governing agency.

State regulatory agencies may also develop notification levels, health advisory levels, or action levels for many micropollutants that do not have USEPA MCLs. For example, 1,4-dioxane, which is primarily treated with AOTs, has a USEPA health advisory level of 0.35 mg/L but does not have a drinking water MCL. Many states have adopted drinking water guidelines for 1,4-dioxane (Table 1-2). As with federally regulated contaminants, compliance with any US state notification or action levels would be based on maintaining concentrations below the required level based on the state required monitoring frequency.

The USEPA also regulates several potential by-products from AOTs as part of the Stage 1 and Stage 2 Disinfectant and Disinfection By-product Rules. These include bromate (ozone-based AOTs and UV/chlorine only) and disinfection by-products (DBPs) (total trihalomethanes and haloacetic acids). AOTs do not directly form regulated DBPs; however, depending on the site, specific water quality can increase the formation potential once chlorine is added to the water. AOT by-products are discussed in more detail in chapter 6.

In the United States, requirements for reuse or recycled water applications are typically more state-specific as many micropollutants found in wastewater are unregulated by the USEPA. Water reuse is defined as "the use of treated municipal wastewater (reclaimed water)" (USEPA et al. 2012) and is synonymous with water recycling. Reuse

Introduction 5

Table 1-2 Example US state guidelines for 1,4-dioxane (as of June 2015)

State	Guideline	Description	Concentration ($\mu g/L$)
California	Notification Level	Health based advisory level with public notification requirements	1
Colorado	Interim Ground Water Standard	State groundwater regulation	0.35
Connecticut	Action Level	Level above which remedial action is recommended	3
Maine	Maximum Exposure Guideline	Non-enforceable recommendation for maximum concentrations	4
Massachusetts	Guideline	Non-regulatory limit set by state	0.3
North Carolina	Groundwater Quality Standard	State maximum allowable concentration	3

can be further divided into potable and nonpotable application. Potable reuse applications are ones where the treated water would be used for augmentation of drinking water supplies either directly or indirectly. Potable reuse applications often employ AOTs for treatment of a broad range of contaminants.

Nine US states (as of 2012) have developed potable reuse rules, regulations, or guidelines for the design and operation of a reuse facility (USEPA et al. 2012). In 2012, the USEPA released the *Guidelines for Water Reuse* (USEPA et al. 2012), which provides general recommendations for treatment and water quality goals for wastewater reuse (Table 1-3). Advanced wastewater treatment, which includes advanced oxidation, is listed as a suggested treatment technology for indirect potable reuse with finished water quality recommended to meet drinking water standards. The reuse guidelines also recommend providing a multibarrier approach for microbiological and chemical contaminants and advanced technologies that address a broader variety of contaminants with greater reliability. AOTs can help to meet both of these recommendations. AOTs may also be useful in nonpotable reuse applications as well, depending on project goals and requirements.

ADVANTAGES AND DISADVANTAGES OF ADVANCED OXIDATION

Advantages and disadvantages for AOTs are described in the following section.

AOT Advantages

1. Degrades organic contaminants and does not transfer to another phase that must be treated or disposed of [e.g., reverse osmosis (RO) brine or granular activated carbon (GAC) media].

2. Converts most refractory organic contaminants (not treatable biologically) into forms that are biologically treatable.
3. Very effective in treating most micropollutants.
4. Nonselective and can treat a broad range of contaminants simultaneously.
5. Oxidizes taste-and-odor compounds from drinking water.
6. Complements other technologies in water reuse applications by providing a barrier for a broad group of microconstituents.
7. Reactions are very fast and contact times are in the range of a few seconds.
8. Provides simultaneous microbial disinfection while degrading contaminants.

AOT Disadvantages

1. Can be expensive in terms of capital and annual operations and maintenance (O&M) costs (i.e., power and chemical requirements).
2. Treated water should be tested for potential regulated and unregulated by-products.
3. Any hydrogen peroxide residual after treatment must be quenched if the water goes to a potable distribution system.
4. Most UV reactors contain mercury lamps, so breakage of UV lamps presents a possible mercury hazard. Site-specific evaluations can be completed to determine potential concentrations.
5. Power interruptions could inhibit effective system operation without expensive power conditioning equipment. This could result in some water not being treated unless the water is diverted to the influent of AOT process or waste.

Table 1-3 USEPA suggested guidelines for water reuse (USEPA et al. 2012)

Indirect Potable Reuse Category	Treatment	Reclaimed Water Quality
Groundwater recharge by injection into potable aquifers or Augmentation of surface water supply reservoirs	• Secondary • Filtration • Disinfection • Advanced wastewater treatment	Includes, but not limited to, the following: • No detectable total coliform/100 mL • 1 mg/L Cl residual, minimum • pH = 6.5–8.5 • ≤2 NTU • ≤2 mg/L TOC of wastewater origin • Meets drinking water standards

AOT HANDBOOK ORGANIZATION

This handbook consists of twelve chapters and three appendices designed to provide basic technology background and important design and operational considerations for AOT. The handbook is organized as follows:

- **Chapter 2—Fundamentals of UV Light and Photochemistry.** Describes the fundamental principles of UV light generation and photochemical reactions as the basis of UV-based AOTs.

- **Chapter 3—Fundamentals of Advanced Oxidation.** Discusses radical formation, treatment mechanisms, and examples of AOTs applications.

- **Chapter 4—Advanced Oxidation Types.** Discusses the available AOTs and the general applications of each technology.

- **Chapter 5—Advanced Oxidation Equipment.** Summarizes the equipment required for the two most commonly used AOT systems in municipal applications (UV- and ozone-based AOTs)

- **Chapter 6—Effects of Water Quality on AOT Systems.** Discusses water quality parameters and by-products that should be considered when designing an AOT system.

- **Chapter 7—Possible Locations for AOT Facilities.** Discusses general applications where AOTs are an applicable technology.

- **Chapter 8—AOT System Design Considerations.** Discusses planning for AOT systems, including treatment goals, basic design parameters, hydraulics, electrical power considerations, treatability testing, and cost evaluations.

- **Chapter 9—Startup, Operations, and Maintenance.** Discusses startup and operation issues for AOT facilities, recommended monitoring, and maintenance tasks.

- **Chapter 10—AOT Case Studies.** Presents examples of AOT applications.

- **Chapter 11—Safety and Handling of AOT Equipment.** Discusses important health and safety concerns when designing and operating an AOT facility.

- **Chapter 12—Considerations for a Water Utility Manager.** Discusses important considerations for utilities considering the design of an AOT facility.

- **Appendix A**   Terms, Units, Symbols, and Definitions.

- **Appendix B**   Rate Constants and Quantum Yields.

- **Appendix C**   Calculation of Fraction of UV Absorbed for UV/H_2O_2 AOT.

REFERENCES

Baxendale, J.H., and J.A. Wilson. 1957. The Photolysis of Hydrogen Peroxide at High Light Intensities. *Trans. Faraday Soc.*, 53:344–356.

Bircher, K., M. Vuong, B. Crawford, M. Heath, and J. Bandy. 2012. Using UV Dose Response for Scale Up of UV/AOP Reactors. In *Proc. of the International Water Association World Congress*, Busan, Korea.

Bolton, J.R. 2010. *Ultraviolet Applications Handbook*, 3rd Ed. Edmonton, AB, Canada: ICC Lifelong Learn Inc.

Bolton, J.R., and S.R. Cater. 1994. Homogeneous Photodegradation of Pollutants in Contaminated Water: An Introduction, In *Aquatic and Surface Photochemistry* Chapter 33, pp. 467-490. G.R. Helz, R.G. Zepp, and D.G. Crosby, Eds. Boca Raton, Fla.: CRC Press,

Bolton, J.R., and C.A. Cotton. 2008. *The Ultraviolet Disinfection Handbook*. Denver, CO: American Water Works Association.

Bolton, J.R., K.G. Bircher, W. Tumas, and C.A. Tolman. 1996. Figures-of-Merit for the Technical Development and Application of Advanced Oxidation Processes. *Jour. Advan. Oxid. Technol.* 1:13–17.

Bolton, J.R., K.G. Bircher, W. Tumas, and C.A. Tolman. 2001. Figures-of-Merit for the Technical Development and Application of Advanced Oxidation Technologies for Both Electric- and Solar-Driven Systems. *Pure Appl. Chem.*, 73(4):627–637.

Braun, A.M., M.-T. Maurette, and E. Oliveros. 1991. *Photochemical Technology*. Chichester, UK: Wiley.

Fenton, H.J.H. 1894. Oxidation of Tartaric Acid in Presence of Iron. *Jour. Chem. Soc.*, 65:899–910.

Getoff, N., and G.O. Schenck. 1968. Primary Products of Liquid Water Photolysis at 1236, 1470, and 1849 Å. *Photochem. Photobiol.*, 8:167–178.

Glaze, W.H., J.W. Kang, and D.H. Chapin. 1987. The Chemistry of Water Treatment Processes Involving Ozone, Hydrogen Peroxide and Ultraviolet Radiation. *Ozone Sci. Eng.*, 9: 335–352.

Hoigné, J., and H. Bader. 1975. Ozonation of Water: Role of Hydroxyl Radicals as Oxidizing Intermediates. *Science*, 190:782–784.

Ikehata, K., and M. Gamal El-Din. 2006. Aqueous Pesticide Degradation by Hydrogen Peroxide/Ultraviolet Irradiation and Fenton-Type Advanced Oxidation Processes: A Review. *Jour. Environ. Eng. Sci.*, 6:81–135.

Kistiakowsky, W. 1900. Experiments on the Sensitiveness to Light of Hydrogen Peroxide in Aqueous Solutions on Addition of Potassium Ferro- and Ferricyanide. *Zeit. Phys. Chim.*, 35:431–439.

Legrini, O., E. Oliveros, and A.M. Braun. 1993. Photochemical Processes for Water Treatment. *Chem. Rev.*, 93:671–698.

Linden, K., U. Gunten, H. Mestankova, and A.A. Parker. 2015. *Advanced Oxidation and Transformation of Organic Contaminants*. Denver, CO: Water Research Foundation.

Nakayama, S., K. Esaki, K. Namba, Y. Taniguchi, and N. Tabata. 1979. Improved Ozonation in Aqueous Systems. *Ozone Sci. Eng.*, 1-2:119–131.

Oppenländer, T. 2003. *Photochemical Purification of Water and Air.* Weinheim, Germany: Wiley-VCH.

Parsons, S. Ed. 2004. *Advanced Oxidation Processes for Water and Wastewater Treatment*. London: IWA Publishing.

Peyton, G.R., F.Y. Huang, J.L. Burleson, and W.H. Glaze. 1982. Destruction of Pollutants in Water With Ozone in Combination With Ultraviolet Radiation. 1. General Principles and Oxidation of Tetrachloroethylene. *Environ. Sci. Technol.*, 16:448–453.

Tarr, M.A. Ed. 2003. *Chemical Degradation Methods for Wastes and Pollutants—Environmental and Industrial Applications*. New York: Marcel Dekker; e-edition, Taylor and Francis.

Taube, H. 1956. Photochemical Reactions of Ozone in Solution. *Trans. Faraday Soc.*, 53:656–665.

United States Environmental Protection Agency (USEPA). 2006. *Ultraviolet Disinfection Guidance Manual for the Final Long Term 2 Enhanced Surface Water Treatment Rule*. Washington, DC: US Environmental Protection Agency, Office of Safe Water. Available on the web at: http://www.epa.gov/safewater/disinfection/lt2/pdfs/guide_lt2_uvguidance.pdf.

USEPA, National Risk Management Research Laboratory, and US Agency for International Development. *2012 Guidelines for Water Reuse*. Washington DC: Office of Wastewater Management and Office of Water.

Urey, H.C., L.H. Dawsey, and F.O. Rice. 1929. The Absorption Spectrum and Decomposition of Hydrogen Peroxide by Light. *Jour. Am. Chem. Soc.*, 51:1371–1383.

Wayne, C.E. and R.P. Wayne. 1996. *Photochemistry*. Oxford, UK: Oxford University Press.

2

Fundamentals of Ultraviolet Light and Photochemistry[1]

Many AOTs are driven by photochemical processes, in which ultraviolet (UV) or visible light[2] drives the formation of highly reactive intermediates, such as hydroxyl radicals. Hence, a proper understanding of AOTs requires some background on the fundamentals of UV, visible light, and photochemistry. This chapter and the next are designed to help the reader understand these fundamental concepts, which are discussed in general terms. Terms and symbols will often be introduced without explanation. Specific definitions of terms, symbols, and units are given in appendix A.

Light is essential to most life forms. Humans see only a very small fraction of the *colors* of light. Except for some AOTs that are driven by visible light, this book primarily addresses light with wavelengths *beyond the violet* end of the rainbow or the *ultraviolet*. There are several references for information on light and its measurement; for example, a free handbook is available from International Light (Ryer 1997). The fundamentals of photochemistry are covered very well in the books by Wayne (1988) and by Wayne and Wayne (1996). The book by Turro et al. (2009) provides a more in-depth coverage of photochemistry. The classic book on photochemistry by Calvert and Pitts (1966) is no longer in print, but it is available in many university libraries. A review by Pfoertner and Oppenländer (2012) provides a good introduction to photochemistry and industrial applications, and the book by Braun et al. (1991) covers practical applications of photochemistry.

---

1 Some of this chapter has been adapted (with permission) from Bolton (2010) and from Bolton and Cotton (2008).
2 Physicists prefer to use the term *ultraviolet radiation*, reserving the term *light* for *visible light*. The authors prefer to use a broader definition of *light* to include the UV, visible, and IR spectral regions because of the negative public connotation for the term *radiation*.

LIGHT AND PHOTONS

Light is a form of electromagnetic rays, an energy form that extends from radio waves to cosmic rays over at least 16 orders of magnitude. However, one can see only a very small fraction of the *colors* of light in the *visible* range.

Traditionally light is seen as having wave properties, such as diffraction and interference. However, Einstein (1905) interpreted Lenard's (1902) photoelectric experiments in terms of light consisting of a stream of particles called *photons*, each with an energy inversely proportional to the wavelength of the light. The relationship between the "wave nature" and the "particle nature" of light is embodied in the Planck Law.

$$u = h\nu = hc/\lambda = hc\bar{\nu} \quad \text{(Eq. 2-1a)}$$

$$U = N_A h\nu = hcN_A/\lambda = hcN_A\bar{\nu} \quad \text{(Eq. 2-1b)}$$

where u is the *energy* (J) of one photon, ν is the *frequency* (s^{-1}), λ is the *wavelength* (m), $\bar{\nu}$ is the *wavenumber* (m^{-1}), c is the *speed of light* (2.9979×10^8 m s^{-1}) in a vacuum, h is the *Planck constant* (6.6261×10^{-34} J s), N_A is the *Avogadro number* (6.02214×10^{23} mol^{-1}), and U is the *energy per einstein*.[3] The units here have been given in the standard Système Internationale (SI) forms (see appendix A); however, for applications in ultraviolet light and photochemistry, λ is usually given in nanometers (nm) and $\bar{\nu}$ in cm^{-1}, with appropriate numerical factors to make Eq. 2-1b equal joules per einstein. Various physical constants are collected in appendix A.

The absorption of photons by a molecule raises that molecule to an excited state from which a chemical reaction may occur (this process is called *photochemistry*—see following sections).

UV LIGHT SPECTRAL RANGES

Photochemistry can occur anywhere in the wavelength range of 100–1,000 nm. Light photons with wavelengths longer than 1,000 nm have photon energies that are too small to cause chemical change when absorbed. Photons with wavelengths shorter than 100 nm have so much energy that molecules are ionized producing electrons and reactive radicals (e.g., ·OH), which cause molecular disruptions characteristic of the effects of radiation from radioactive sources. The total photochemical wavelength range is divided into bands with specific names as given in Figure 2-1 and Table 2-1.

Little photochemistry occurs in the *near infrared* (700–1,000 nm), except for some photosynthetic bacteria, which are capable of storing solar energy at wavelengths up to almost 1,000 nm. The *visible* range (400–700 nm) is completely active for photosynthesis in green plants and algae. Also, many dyes undergo photochemical transformations in the visible range or sensitize reactions in other molecules.

---

3 An *einstein* is 1 mole (6.0221418×10^{23}) of photons.

Fundamentals of Ultraviolet Light and Photochemistry

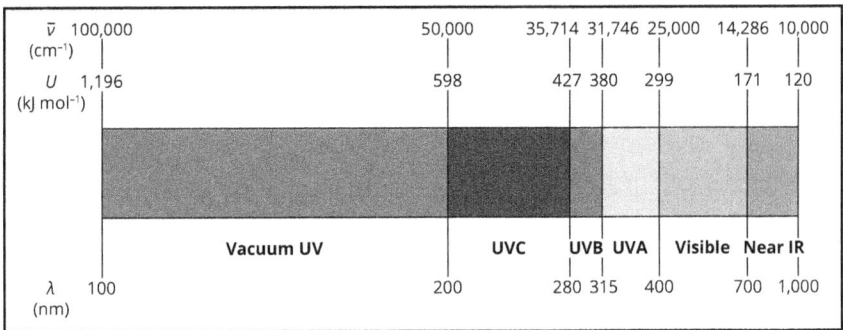

Figure 2-1 Spectral ranges of UV, visible, and IR light. The ultraviolet range is 100–400 nm.

Table 2-1 Spectral ranges of interest in photochemistry (see Figure 2-1)

Range Name	Wavelength Range (nm)	Wavenumber Range (cm^{-1})	Energy Range (kJ einstein^{-1})
Near Infrared	700–1,000	10,000–14,286	120–171
Visible	400–700	14,286–25,000	171–299
Ultraviolet UVA	315–400	25,000–31,746	299–380
UVB	280–315	31,746–35,714	380–427
UVC	200–280	35,714–50,000	427–598
Vacuum UV (VUV)	100–200	50,000–100,000	598–1,196

Most studies in photochemistry involve the *ultraviolet* ranges (100–400 nm). This region is divided into four subranges connected with the human skin's sensitivity to ultraviolet light.

1. The UVA range (315–400 nm) causes changes in the skin that lead to sun tanning.
2. The UVB range (280–315 nm) can cause sun burning and eventually can induce skin cancer.
3. The UVC range (200–280 nm) is absorbed by RNA, DNA, and proteins and can lead to cell mutations, cancer, or cell death.
4. The VUV range is absorbed by almost all substances (including water and air). Thus VUV can only be transmitted in a vacuum. The absorption of a VUV photon causes one or more bond breaks. For example, in the gas phase oxygen (O_2) absorbs VUV as follows to form ozone (O_3).

$$O_2 + \text{VUV photon} \rightarrow 2\ O(^1D) \qquad \text{(Eq. 2-2a)}$$

O(^{1}D) is an excited state of the oxygen atom and rapidly converts to the ground state O(^{3}P) atom, which then reacts with O$_2$ as follows:

$$O(^3P) + O_2 + M \rightarrow O_3 + M^* \qquad \text{(Eq. 2-2b)}$$

where M is another molecule (e.g., O$_2$ or N$_2$). M is necessary to remove some kinetic energy to allow the O$_3$ molecule to form.

With regards to humans, UVC must be treated with appropriate care. UVC is absorbed strongly in the surface layer of skin cells, which are sloughed off almost daily. Therefore, moderate exposures can be tolerated. However, UVC exposure to the eyes is more dangerous because it can cause the formation of cataracts (Wieringa 2006).

The UVC range is sometimes called the *germicidal range* because it is very effective in inactivating bacteria and viruses. However, light-driven AOTs can involve wavelengths from the vacuum UV to the visible range, although most commercial AOTs involve photochemistry in the 200–300 nm range.

EMISSION, TRANSMISSION, AND ABSORPTION OF LIGHT

The various aspects of light can be discussed in the context of *emission* from a UV lamp, *transmission* through a medium (air, water, etc.), and *absorption* by a target molecule.

Light Transmission and Absorption

Light is transmitted at a speed of 2.9979 × 10^8 m s^{-1} in a vacuum but slows down when entering any finite medium (e.g., air or water). The ratio of the speed of light in a vacuum to that in a given medium is called the *refractive index*, the value of which is 1.0000 in a vacuum [the refractive index of air is very close to unity (1.0003)]. When light passes from one medium of refractive index n_1 to another with refractive index n_2 ($n_2 > n_1$), two effects occur.

1. Some of the light is *reflected* back from the interface[4] such that the angle of reflection (θ_r) is equal to the angle of incidence (θ_1). For light directly incident on a surface, the reflection coefficient is given by

$$R = \frac{(n_2 - n_1)^2}{(n_2 + n_1)^2} \qquad \text{(Eq. 2-3)}$$

If the two media are air and water, $R = 0.025$ at 254 nm.

---

4 This is the familiar phenomenon of seeing the *reflection* of a mountain on the surface of a lake. The amount of reflection depends on the angle of incidence, with more light being reflected for more oblique angles.

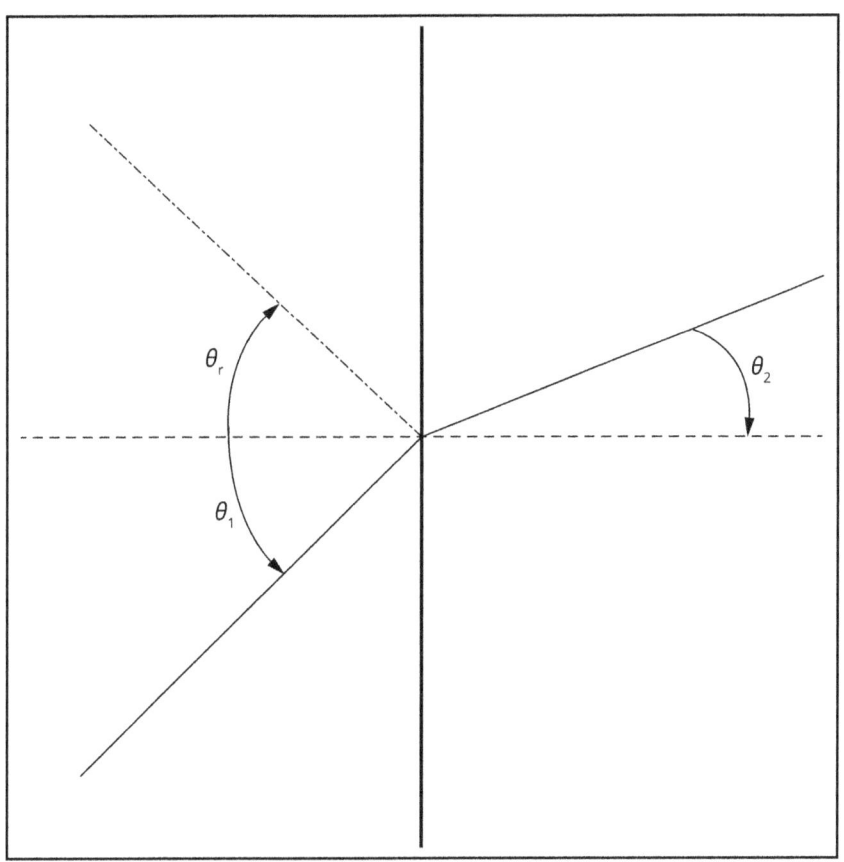

Figure 2-2 Reflection and refraction as a light beam passes from medium 1 with refractive index n_1 to medium 2 with refractive index n_2. The dotted line represents the reflected light. θ_1 is the angle of incidence, θ_r (= θ_1) is the angle of reflection, and θ_2 (< θ_1) is the angle of refraction.

2. Light entering the new medium is *refracted* so that its direction is changed; that is, the angle to the normal at the surface (θ_2) is less than the angle of incidence (θ_1). The magnitude of these two effects increases as $n_2 - n_1$ becomes larger. These effects are illustrated in Figure 2-2.[5]

If there are absorbing substances in the medium, the light is attenuated as it passes through according to the Beer-Lambert Law.

$$\frac{E'(\lambda)}{E^0(\lambda)} = 10^{-A(\lambda)} = T(\lambda) \qquad \text{(Eq. 2-4a)}$$

---

[5] For a quantitative discussion of reflection and refraction, see Bolton (2010).

or

$$A(\lambda) = \log\left(\frac{E^0(\lambda)}{E^t(\lambda)}\right) = -\log[T(\lambda)] = a(\lambda)l \qquad \text{(Eq. 2-4b)}$$

$E^t(\lambda)$ and $E^0(\lambda)$[6] are the transmitted and incident irradiances at wavelength λ as a beam passes through a medium over a path length l; $A(\lambda)$ is the *absorbance* (unitless) at wavelength λ, and $T(\lambda)$ is the *transmittance* (unitless) at wavelength λ.[7] $A(\lambda)$ is directly proportional to the path length l, that is, $A(\lambda) = a(\lambda)l$, where $a(\lambda)$ is the *absorption coefficient* (cm^{-1}) at wavelength λ. It is important to understand that Eq. 2-4a and 2-4b apply only for a beam of light with a narrow range of wavelengths. In fact, a plot of $A(\lambda)$ versus wavelength is called the *absorption* or *absorbance spectrum*, and a plot of $T(\lambda)$ versus wavelength is called the *transmittance spectrum*.

Eq. 2-4a and 2-4b assume that no scattering occurs in the beam. If scattering by particles is present, the situation is more complex because some of the light is scattered at oblique angles and some may be absorbed by the particles.

The absorbance is related to the concentrations of absorbing components by

$$A(\lambda) = \sum_i \varepsilon_i(\lambda) c_i l \qquad \text{(Eq. 2-5)}$$

where $\varepsilon_i(\lambda)$ is the *molar absorption coefficient* (M^{-1} cm^{-1})[8] at wavelength λ for component i; c_i is the *concentration* (M) of component i in the solution; and l is the path length (cm). Eq. 2-5 is employed in measurements of absorbance that are often used to determine concentrations.

Often the transmittance of UV light in the medium (e.g., drinking water) is described by the ultraviolet transmittance (UVT), which is defined as the percent transmittance in the medium when the path length is 1.0 cm and the wavelength is 254 nm. That is

$$\text{UVT} = 100T(254) = 100 \times 10^{-A(254)} \qquad \text{(Eq. 2-6)}$$

where $A(254)$ is the absorbance[9] at 254 nm in a 1.0 cm path length.

Absorbance and transmittance are measured with a spectrophotometer, which measures the irradiance of UV light as it passes through a quartz cell containing the solution of interest (e.g., a drinking water sample). Because a small fraction of the UV

---

6 When a monochromatic light beam is used, the absorbance is often given as just A and the transmittance as just T; however, it should be remembered that A and T are wavelength dependent.
7 Often the transmittance is expressed as the *percent transmittance* (%T), which usually implies a 1 cm path length.
8 The SI units for the molar absorption coefficient are m^2 mol^{-1}; 1 m^2 mol^{-1} = 10 M^{-1} cm^{-1}.
9 This is really an *absorption coefficient* because the path length is defined as 1 cm.

beam is reflected from the quartz surfaces, a *blank* measurement must be made with only pure solvent (e.g., distilled water) in the quartz cell. Thus in practice, the ratio E^t/E^0 in Eq. 2-4a is the ratio of the detector level for the cell with the solution of interest to that for the cell containing pure solvent. The *transmittance* is displayed (or plotted) as this ratio itself. In the case of *absorbance*, the spectrophotometer performs the negative logarithm of this ratio. Because many UV reactors use low-pressure lamps, which primarily emit at 254 nm, the measurement of the absorbance at 254 nm [$A(254)$] has become a standard water quality parameter, even when medium-pressure UV lamps are used. There exists a standard method for $A(254)$ (APHA, AWWA, and WEF 2012). When using the Standard Method for measuring $A(254)$ for AOT applications, samples should not be filtered prior to analysis as the method recommends.

Most drinking waters have UVT values between 75–99 percent. This means that E^t is almost the same as E^0. Significant errors can occur if one uses a 1.0 cm path length cell for highly transmitting waters. To minimize these errors, a 5.0-cm or 10.0-cm cell should be used. $A(254)$ is the measured absorbance divided by the path length.

EXCITED STATES AND THE ABSORPTION PROCESS

When light is absorbed in a medium, it causes changes in the electronic state of components in the medium. There are two types of absorption: discrete *molecular* absorption and delocalized *band* absorption (see Figure 2-3).

Excited-State Processes

In the case of molecules dissolved in a dilute solution (e.g., aqueous solution) or in a gas, the absorbing molecules can be considered to be discrete, independent absorbing entities. Absorption of a photon (with sufficient energy) causes a change in the electronic state of the molecule from a singlet[10] ground state (S_0) to one or higher excited singlet states ($S_1, S_2, \ldots$).

The excited singlet state can then undergo several processes.

1. Any molecules that are excited to S_2 or higher excited singlet states relax very quickly (<10 ps) to the S_1 state via internal conversion (IC) where the excess energy is released as heat. This is called *thermalization* of the excited state.

2. The S_1 state can return to the ground state either by emission of a photon (hv_f, fluorescence) or by converting the excess energy to heat [internal conversion (IC)].

---

[10] The designation of *singlet* or *triplet* refers to the electronic configuration, that is, the *multiplicity* of the electronic spins in the state. Molecules in which all electron spins are paired are in a *singlet* state. If the molecule has two unpaired electrons, the molecule is in a *triplet* state. Molecules that have an odd number of electrons (e.g., NO) usually have one unpaired electron and are in a *doublet* state.

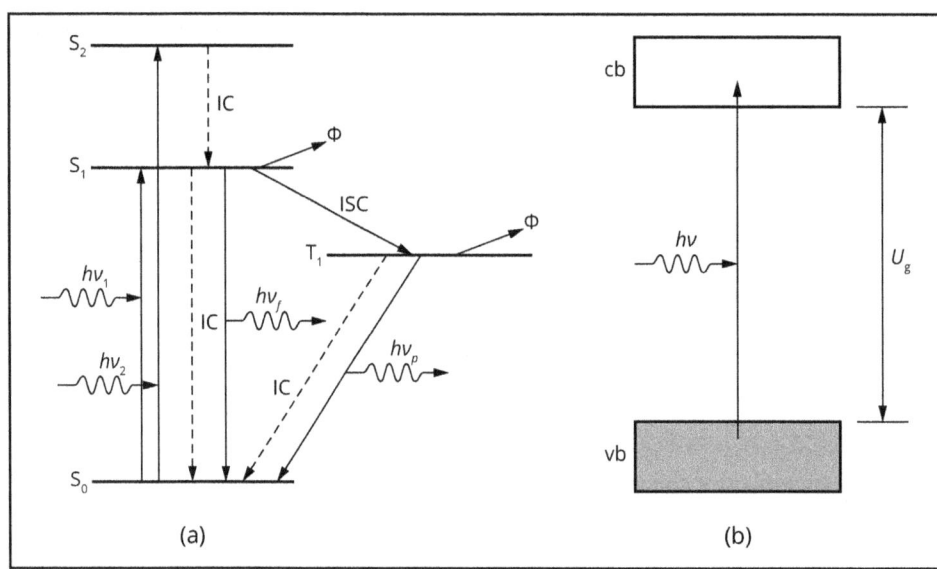

Figure 2-3 (a) Discrete molecular absorption: molecular energy level diagram showing excitation from the ground S_0 singlet state to either the S_1 or S_2 excited singlet state. T_1 is the lowest triplet state. The dotted lines indicate nonradiative decay, called *internal conversion* (IC). The process of converting a singlet state to a triplet state is called *intersystem crossing* (ISC). And (b) Delocalized "band" absorption (as in a semiconductor): electronic transition in a semiconductor from the lower *valance band* (vb) to the upper *conduction band* (cb). U_g is the *bandgap energy*. Photochemistry can occur from either the state S_1 or the state T_1 with distinct quantum yields (Φ).

3. S_1 can transform into a *metastable* long-lived triplet excited state [$S_1 \rightarrow T_1$ conversion, called *intersystem crossing* (ISC)], which can then emit a photon (hv_p, phosphorescence) or convert the excess energy to heat by IC.

4. S_1 or T_1 can undergo chemical reaction (i.e., photochemistry) with distinct quantum yields (Φ). Photochemistry involves breaking or rearrangement of chemical bonds in the molecule.

5. In the case of a delocalized solid matrix, such as a semiconductor, absorption of a photon causes an electron to be transferred from a *valance band* (vb) (almost full of electrons) to a *conduction band* (cb) (almost empty of electrons) across a *bandgap* U_g, which has virtually no energy levels. This creates an excess *electron* in the cb and a *hole* in the vb.

Quantum Yield

The *quantum yield* (unitless) Φ is a measure of the photon efficiency of a photochemical reaction. Φ is defined as the number of moles of reactant (R) removed or product (P) formed per einstein of photons absorbed. If a steady state is achieved, Φ can also be defined in terms of rates as

$$\Phi_R = \frac{\text{rate of removal of R}}{\text{rate of absorption of photons}} \quad \text{(Eq. 2-7a)}$$

$$\Phi_P = \frac{\text{rate of generation of P}}{\text{rate of absorption of photons}} \quad \text{(Eq. 2-7b)}$$

where the rate of generation of P (or rate of removal of R) is in mol s^{-1} and the rate of absorption of photons is the absorbed photon flow (einstein s^{-1}). Note that Φ is based on the *absorbed* and not the *incident* photon flow.

LAWS OF PHOTOCHEMISTRY

Because photochemistry is driven by the absorption of photons, three *laws* have been developed to describe the factors that limit photochemical processes.

First Law of Photochemistry

The first law of photochemistry as stated by T. Grotthuss and J.W. Draper more than 150 years ago is

> *Only the light, which is absorbed by a molecule, can be effective in producing photochemical change in the molecule.*

This means that molecules that do not absorb light in a given range of wavelengths cannot undergo a photochemical reaction when exposed to light in those wavelengths.

Second Law of Photochemistry

The second law of photochemistry deduced by J. Stark and A. Einstein about 100 years ago states that

> *For each photon of light absorbed by a chemical system, only one molecule is activated for a photochemical reaction.*

This law implies that each photon of light can (at most) cause the photochemical reaction of just one light-absorbing molecule.

A corollary of this law states that the total amount of photoreaction that occurs is directly proportional to the product of the absorbed photon flow and the time of illumination. That is, more light produces more photoproduct,[11] or 1 mole of photons absorbed

---

[11] This law breaks down for very high photon flows, such as from a very powerful laser. In this case, the concentration of excited states is high enough that excited states themselves can absorb a photon, and a process called *multiphoton absorption* is possible. Fortunately, this condition is almost never encountered in ultraviolet reactors.

by 1 mole of reactant will lead to 1 mole of photoproduct if the quantum yield Φ of the process is unity.

A second corollary is that all photochemical events are independent of each other.

Third Law of Photochemistry

The energy of an absorbed photon must be equal to or greater than the weakest bond in the molecule.[12]

A photochemical reaction usually leads to the breaking of one or more bonds in the molecule. From the law of conservation of energy, the energy of the absorbed photon must be at least as large as that of the weakest bond broken.

As an example, by absorption of UV light, the photolysis of nitric oxide (NO_2), found in the smog over many cities in the summer, leads to breaking the N–O bond in NO_2. But this bond has a high energy (305 kJ mol^{-1}), so that even though NO_2 absorbs light out to 700 nm, only UV light with wavelengths less than 392 nm can lead to the photochemical breakup of NO_2. Light of longer wavelengths is absorbed, but the excited state simply returns to the ground state releasing the energy as heat.

REFERENCES

APHA, AWWA, and WEF. 2012. *Standard Methods for the Examination of Water and Wastewater*, 22nd Ed. Washington, DC: APHA, AWWA and WEF.

Bolton, J.R., 2010. *Ultraviolet Applications Handbook*, 3rd Ed. Edmonton, AB, Canada: ICC Lifelong Learn Inc.

Bolton, J.R. and C.A. Cotton. 2008. *The Ultraviolet Disinfection Handbook*. Denver, CO: American Water Works Association.

Braun, A.M., M.-Y. Maurette, and E. Oliveros. 1991. *Photochemical Technology*. Chichester, UK: Wiley.

Calvert, J.G., and J.N. Pitts Jr. 1966. *Photochemistry*. New York: Wiley.

Einstein, A. 1905. Über einen die Erzeugung und Verwandlung des Lichtes betreffenden huerestischen Gesichtspunkt (On a Heuristic Point of View Concerning the Production and Transformation of Light). *Ann. Phys.*, 17:132–148.

Lenard, P. 1902. Über die Lichtelektrische Wirkung (Concerning the Photoelectric Effect). *Ann. Phys.*, 8:149–198.

Pfoertner, K.-H., and T. Oppenländer. 2012. *Photochemistry in Ullmann's Encyclopedia of Industrial Chemistry*. Verlag, Weinheim, Germany: Wiley-VCH.

Ryer, A. 1997. *The Light Measurement Handbook*. Newburyport, MA: International Light, http://www.intl-lighttech.com/services/light-measurement-handbook.

Turro, N.J., V. Ramamurthy, and J.C. Scaiano. 2009. *Principles of Molecular Photochemistry, an Introduction*. Sausalito, CA: University Science Books.

---

12 This law does not apply to photochemical reactions in which bonds are rearranged, such as in a *cis–trans* conversion in a molecule with double bond(s).

USEPA. 2006. *Ultraviolet Disinfection Guidance Manual for the Final Long Term 2 Enhanced Surface Water Treatment Rule*. Washington, DC: US Environmental Protection Agency. http://www.epa.gov/safewater/disinfection/lt2/pdfs/guide_lt2_uvguidance.pdf.

Wayne, R.P. 1988. *Principles and Applications of Photochemistry*. Oxford, UK: Oxford University Press.

Wayne, C.E., and R.P. Wayne. 1996. *Photochemistry*. Oxford, UK: Oxford University Press.

Wieringa, F.P. 2006. "Five Frequently Asked Questions About UV Safety." *IUVA News* 8(2):29–32.

3

Fundamentals of Advanced Oxidation[1]

In this chapter, fundamental photochemical and kinetic reactions are explained in relation to advanced oxidation technology (AOT) processes. Because the vast majority of AOTs involve the generation and reaction of hydroxyl radicals, this chapter begins with a description of the properties of the hydroxyl radical.

Some of the material presented in this chapter is rather complex and involves many equations. These sections have been highlighted in grey. Readers may skip these sections unless they are interested in carrying out photochemical rate studies.

See appendix A for definitions and symbols for the terms used in this chapter.

PROPERTIES OF THE HYDROXYL RADICAL

The hydroxyl radical (·OH) is one of the most powerful oxidizing agents known in aqueous systems ($E° = +2.7$ V in acid solution) and reacts with most organic compounds at almost diffusion-controlled rates, which results in rapid oxidation kinetics. For comparison, free chlorine (Cl_2), which is a strong oxidizer commonly used in water treatment, has an $E°$ of +1.36 V.

The following are possible hydroxyl radical reaction types:

1. *Abstraction of a hydrogen atom*—usually with aliphatic hydrocarbon groups; for example

$$CH_3OH + ·OH \rightarrow ·CH_2OH + H_2O \qquad \text{(Eq. 3-1a)}$$

2. *Addition* – usually with unsaturated or aromatic hydrocarbon groups; for example

$$CHCl = CCl_2 (TCE) + ·OH \rightarrow ·CHCl - C(OH)Cl_2 \qquad \text{(Eq. 3-1b)}$$

---

[1] Some of this chapter has been adapted (with permission) from Bolton (2010).

3. *Electron transfer*—usually with inorganics; for example

$$\text{I}^- + \cdot\text{OH} \rightarrow \cdot\text{I} + \text{OH}^- \qquad \text{(Eq. 3-1c)}$$

4. *Radical-radical reactions*—for example

$$\text{HO}_2\cdot + \cdot\text{OH} \rightarrow \text{O}_2 + \text{H}_2\text{O} \qquad \text{(Eq. 3-1d)}$$

The following are some key aspects of ·OH radicals that pertain to AOTs:

- ·OH radicals weakly absorb UV light (ε_{230} = 530 M^{-1} cm^{-1}; ε_{260} = 370 M^{-1} cm^{-1}); however, in practice, ·OH radicals absorb very little UV because the steady-state concentration of ·OH radicals is usually below 10^{-9} M.
- After formation, the ·OH radical can dissociate in basic solution (pK_a = 11.9) to form the ·O$^-$ ion.
- The ·OH radical reacts very rapidly with a wide variety of organic and inorganic reagents, often with second-order rate constants approaching the diffusion-controlled limit (~10^{10} M^{-1} sec^{-1}).

Appendix B provides a partial list of rate constants; the Notre Dame Radiation Lab and the National Institute for Science and Technology maintain a useful database of ·OH radical rate constants (NDRL/NIST 2015).

AOT MECHANISM

·OH radicals are powerful oxidants and capable of nonselective oxidation of a variety of organic contaminants. The ·OH oxidizes contaminants through a stepwise degradative reaction mechanism where the products of each step further react in a series of oxidation reactions. The degradation mechanism of organic molecules can be complex, but the general reaction sequence is as follows:

$$\text{pollutant} \rightarrow \text{aldehydes} \rightarrow \text{carboxylic acids} \rightarrow \text{bicarbonate}$$

To illustrate a degradation sequence, ·OH degrades methanol according to the following mechanism:

$$\textbf{CH}_3\textbf{OH} + \cdot\text{OH} \rightarrow \cdot\text{CH}_2\text{OH} + \text{H}_2\text{O} \qquad \text{(Eq. 3-2a)}$$

$$\cdot\text{CH}_2\text{OH} + \text{O}_2 \rightarrow \cdot\text{OOCH}_2\text{OH} \qquad \text{(Eq. 3-2b)}$$

$$\cdot\text{OOCH}_2\text{OH} \rightarrow \textbf{HCHO} + \text{HO}_2\cdot \qquad \text{(Eq. 3-2c)}$$

$$\textbf{HCHO} + \cdot\text{OH} \rightarrow \cdot\text{CHO} + \text{H}_2\text{O} \qquad \text{(Eq. 3-2d)}$$

$$\cdot CHO + O_2 \rightarrow \cdot OOCHO \qquad \text{(Eq. 3-2e)}$$

$$\cdot OOCHO + H_2O \rightarrow \mathbf{HCOOH} + HO_2 \cdot \qquad \text{(Eq. 3-2f)}$$

$$\mathbf{HCOOH} + \cdot OH \rightarrow \cdot COOH + H_2O \qquad \text{(Eq. 3-2g)}$$

$$\cdot COOH + O_2 \rightarrow \cdot OOCOOH \qquad \text{(Eq. 3-2h)}$$

$$\cdot OOCOOH + H_2O \rightarrow \mathbf{H_2CO_3} + HO_2 \cdot \qquad \text{(Eq. 3-2i)}$$

$$2HO_2 \cdot \rightarrow \mathbf{H_2O_2} + O_2 \qquad \text{(Eq. 3-2j)}$$

Note that oxygen (O_2) is almost always consumed in AOT processes.

AOT EXPERIMENTS IN A COLLIMATED BEAM

To further evaluate and optimize AOT processes, contaminant destruction, and by-product formation, AOT processes are often studied in a laboratory setting. Laboratory experiments often use a collimated beam apparatus,[2] such as those shown in Figure 3-1. Collimated beams provide controlled exposures where environmental effects can be accounted for and characterized and UV doses can be precisely measured. This discussion is intended to give a background on collimated beam apparatuses and calculations to benefit later discussions regarding hydroxyl radical reactions.

In the following analysis, equations are often given without derivation. The full derivations can be found in Bolton et al. (2015).

Bolton and Linden (2003) have given a detailed protocol of how to carry out experiments in a collimated beam. This protocol specifies that the average irradiance (or fluence rate)[3] in the water in the dish or beaker placed in the collimated beam is

$$\overline{E}_o(\text{water})(\lambda) \cong \overline{E}(\text{water})(\lambda) = E^0(\lambda) \times \text{PF} \times \text{DF} \times \text{RF}(\lambda) \times \text{WF}(\lambda) \qquad \text{(Eq. 3-3a)}$$

Where
    PF is the petri factor defined as the ratio of the irradiance averaged across the water surface to that in the center of the dish
    DF is the divergence factor; $\text{DF} = \dfrac{D}{D+\ell}$, where D is the distance from the surface of the water to the UV lamp and ℓ is the depth of the solution

---

[2] The collimated beam is actually a *quasi*-collimated beam because the beams are not exactly parallel. However, for the purposes of this treatment, the term *collimated beam* will be used.
[3] In a collimated beam, the irradiance and the fluence rate are virtually identical.

RF(λ) is the reflection factor for water at a given wavelength λ; for λ = 253.7 nm, *RF*(253.7 nm) = 0.975 (Daimon and Masumura 2007); note that *RF*(λ) = [1 − *R*(λ)], where *R*(λ) is the reflection coefficient of water at wavelength. For light incident normal to the surface, $R(\lambda) = [n_1(\lambda) - n_2(\lambda)]^2/[n_1(\lambda) + n_2(\lambda)]^2$, where $n_1(\lambda)$ and $n_2(\lambda)$ are the refractive indices for medium 1 and medium 2 (e.g., *n* = 1.00 for air and *n* = 1.375 for water at 254 nm)

WF(λ) is the water factor at a given wavelength λ. The WF(λ) is derived from an integration of the Beer-Lambert Law from the top to the bottom of the solution. $\text{WF}(\lambda) = \dfrac{1 - 10^{-a(\lambda)l}}{\ln(10)a(\lambda)l}$, where a(λ) is the absorption coefficient (cm^{-1}) of the aqueous solution at wavelength λ, and *l* is the path length (cm), that is, the depth of solution

Alternatively, photochemical reactions can be considered in terms of the photon fluence rate $E_{p,o}$ (einstein m^{-2} s^{-1}) (Bolton et al. 2015), because photochemical reactions are driven by photon flux (Bolton et al. 2015), rather than energy flux. Hence, Eq. 3-3a should be replaced by

$$\overline{E}_{p,o}(\text{water})(\lambda) \cong \overline{E}_p(\text{water})(\lambda) = E_p^0(\lambda) \times \text{PF} \times \text{DF} \times \text{RF}(\lambda) \times \text{WF}(\lambda) \qquad \text{(Eq. 3-3b)}$$

where $E_p^0(\lambda)$ is the incident photon irradiance at wavelength λ. The photon fluence delivered to the water is then

$$F_{p,o}(\text{water})(\lambda) = \overline{E}_p(\text{water})(\lambda)t \qquad \text{(Eq. 3-3c)}$$

Most light-driven AOTs involve sensitized photochemical reactions, in which a primary photochemical process (e.g., the photolysis of H_2O_2) produces a radical intermediate (e.g., the ·OH radical), which then reacts with the target contaminant(s). In the following analysis, the kinetics of sensitized photochemical reactions will be introduced first, followed by an analysis of the kinetics of direct photolysis reactions.

For the purposes of this discussion, we introduce a slightly modified masked collimated beam setup as shown in Fig. 3-2. The purpose of the mask is to ensure that no UV beams strike the walls of the beaker or dish under the mask. Without a mask, some beams can be reflected from the walls back into the solution, which makes the analysis more difficult.

If *D* is the distance from the mask to the lamp; *d* is the distance from the mask to the solution surface; *r* is the inner radius of the beaker; and *x* is the difference between *r* and the radius of the opening in the mask; then this condition requires that $x \geq rd/(D + d)$.

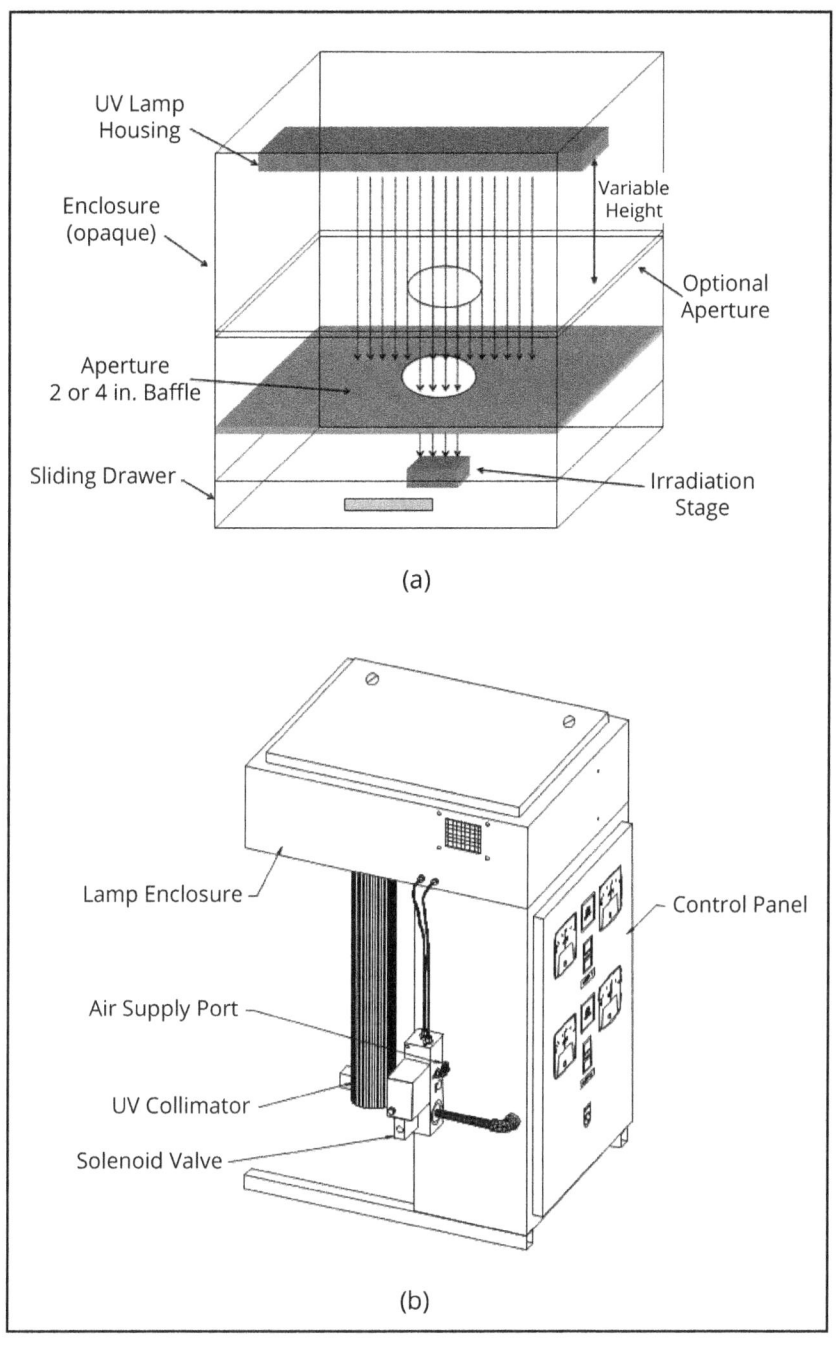

Figure 3-1 (a) CB1 collimated beam apparatus (courtesy of Prof. Karl Linden, private communication) and (b) CB2 commercial collimated beam apparatus (courtesy Calgon Carbon Corporation, Pittsburgh, Pa.)

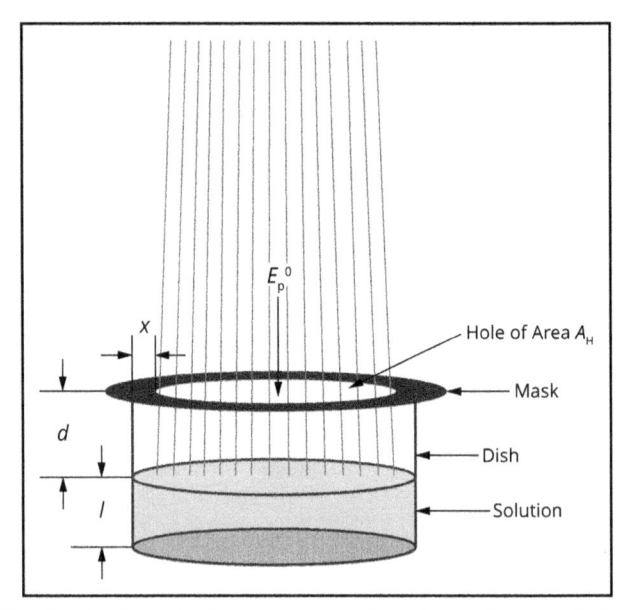

Figure 3-2 Masked collimated beam setup for photochemical studies.

SENSITIZED PHOTOCHEMICAL REACTIONS

A sensitized photochemical reaction is one in which an activator B (e.g., H_2O_2) absorbs light to form a radical R· (e.g., ·OH), which then reacts with a contaminant C to form products. R· may also react with scavengers (possibly including B).

The kinetics of sensitized photochemical reactions can be simplified considerably through the application of the *steady-state approximation*, which sets the rate of formation of highly reactive intermediates (such as R·) to zero.

The following general set of reaction steps apply:

$$B + UV \text{ photons} \rightarrow nR\cdot \qquad R_1 = nR_B \; (M\,s^{-1}) \qquad \text{(Eq. 3-4a)}$$

$$R\cdot + C \rightarrow \text{products} \qquad R_2 = k_C[R\cdot][C] \qquad \text{(Eq. 3-4b)}$$

$$R\cdot + S_i \rightarrow \text{products} \qquad R_3 = \sum_i k_{S_i}[R\cdot][S_i] \qquad \text{(Eq. 3-4c)}$$

where, R_1 is the rate of generation of R· radicals. The S_i represent scavengers that can react rapidly with the R· radicals, and n is the number of R· radicals formed from the photolysis of B. Common scavengers are H_2O_2, HCO_3^-, CO_3^{2-}, and natural organic matter (NOM).

The steady-state approximation sets the rate of formation of R· to zero, therefore

$$\frac{d[R\bullet]}{dt} = nR_B - k_C[C][R\bullet] - \sum_i k_{S_i}[S_i][R\bullet] \cong 0 \qquad \text{(Eq. 3-5a)}$$

or

$$[R\bullet] = \frac{nR_B}{k_C[C] + \sum_i k_{S_i}[S_i]} \qquad \text{(Eq. 3-5b)}$$

The rate of degradation of C is

$$R_2 = -\frac{d[C]}{dt} = k_C[C][R\bullet] = \frac{k_C[C]nR_B}{k_C[C] + \sum_i k_{S_i}[S_i]} = SF\,nR_B \qquad \text{(Eq. 3-5c)}$$

where SF is called the *scavenging factor* given by

$$SF = \frac{k_C[C]}{k_C[C] + \sum_i k_{S_i}[S_i]} \qquad \text{(Eq. 3-5d)}$$

If there are no scavengers, SF = 1.0. Note that if $\sum_i k_{S_i}[S_i] \gg k_C[C]$, Eq. 3-5c becomes linear in [C] and thus is a first-order reaction; whereas if $k_C[C] \gg \sum_i k_{S_i}[S_i]$, Eq. 3-5c becomes independent of [C] and thus is a zero-order reaction. The former is the most common case.

RATES OF DIRECT PHOTOLYSIS REACTIONS WITH MONOCHROMATIC LIGHT

This section evaluates the kinetics of direct photolysis first-order reactions (e.g., reaction Eq. 3-4a) in a collimated beam setup with a monochromatic UV lamp (e.g., a low-pressure UV lamp). Reactions with a polychromatic UV lamp (e.g., a medium-pressure UV lamp) will be evaluated later.

Consider a photochemical reaction of a compound B in a collimated beam setup, such as that shown in Fig. 3-2, in which a monochromatic UV lamp (e.g., low pressure or low-pressure high-output) emitting at one wavelength (e.g., 253.7 nm) is used.

$$B + h\nu \rightarrow \text{products} \qquad \text{(Eq. 3-6)}$$

Rate Expressions

The rate of a direct photolysis reaction depends on three factors—the photon flux, the fraction of the photon flux absorbed by the target compound, and the quantum yield.

The general expression for the rate $R_B(\lambda)$ (M s^{-1}) of reaction Eq. 3-4a under all conditions is the following (Leifer 1988, Bolton et al. 2015):

$$R_B = -\frac{d[B]}{dt} = \frac{q_p(\lambda)\chi_B(\lambda)\Phi_B(\lambda)}{V}$$
$$= \frac{E_p^0(\lambda) \times PF \times RF(\lambda) \times A_H \times \chi_B(\lambda)\Phi_B(\lambda)}{V} \qquad \text{(Eq. 3-7)}$$

where $q_p(\lambda)$ is the photon flux (einstein sec^{-1}) incident on the hole in the mask; $\chi_B(\lambda)$ is the fraction of the photon flux incident into the water that is absorbed by component B; $\Phi_B(\lambda)$ is the quantum yield for the photolysis of B; $E_p^0(\lambda)$ is the incident photon irradiance (einstein m^{-2} s^{-1}) at the center of the hole in the mask at the level of the mask; and RF(λ) is the reflection factor at the air/water surface—all at wavelength λ. PF is the petri factor calculated over the hole in the mask at the level of the mask, A_H is the area (m^2) of the hole in the mask, and V is the volume (L).

$\chi_B(\lambda)$ is the fraction of light incident into the water absorbed by B and is given by

$$\chi_B(\lambda) = \left(\frac{a_B(\lambda)}{a_B(\lambda) + a_{bkgd}(\lambda)}\right)\left(1 - 10^{-[a_B(\lambda) + a_{bkgd}(\lambda)]l}\right) \qquad \text{(Eq. 3-8)}$$

where $a_B(\lambda)$ and $a_{bkgd}(\lambda)$ (cm^{-1}) are the absorption coefficients of B and the background, respectively, and l is the path length (cm).

Consider first the case where $a_{bkgd}(\lambda)l \gg a_B(\lambda)l$; $\chi_B(\lambda)$ in Eq. 3-8 then becomes

$$\chi_B(\lambda) \cong \left(\frac{a_B(\lambda)}{a_{bkgd}(\lambda)}\right)\left(1 - 10^{-a_{bkgd}(\lambda)l}\right)$$
$$= \ln(10)\varepsilon_B(\lambda)l[B]\left[\frac{1 - 10^{-a_{bkgd}(\lambda)l}}{\ln(10)a_{bkgd}(\lambda)l}\right] \qquad \text{(Eq. 3-9)}$$
$$= \ln(10)\varepsilon_B(\lambda)l\,WF(\lambda)[B]$$

where WF(λ) is the water factor for the background absorbance.

Combining Eqs. 3-7 and 3-9

$$-\frac{d[B]}{dt} = \frac{\ln(10)}{10}\frac{E(\lambda)\varepsilon_B(\lambda)RF(\lambda) \times PF \times WF(\lambda)\Phi_B(\lambda)}{U(\lambda)}[B] \qquad \text{(Eq. 3-10a)}$$

The factor 10 is required if $\varepsilon_B(\lambda)$ has units M^{-1} cm^{-1}.

Eq. 3-10a is a first-order rate expression with a first-order rate constant

$$k_1(\lambda) = \frac{\ln(10)}{10} \frac{E(\lambda)\varepsilon_B(\lambda)\text{RF}(\lambda) \times \text{PF} \times \text{WF}(\lambda)\Phi_B(\lambda)}{U(\lambda)} \quad \text{(Eq. 3-10b)}$$

Following Bolton and Linden (2003), Eq. 10b can be written as

$$k_1(\lambda) = \frac{\ln(10)}{10} \frac{\overline{E}(\text{water})(\lambda)\varepsilon_B(\lambda)\Phi_B(\lambda)}{U(\lambda)} \quad \text{(Eq. 3-10c)}$$

where $\overline{E}$(water)(λ) is the volume averaged irradiance in the water column.

The quantum yield can be obtained by rearranging Eq. 3-10c

$$\Phi_B(\lambda) = \frac{10}{\ln(10)} \frac{k_1(\lambda)U(\lambda)}{\overline{E}(\text{water})(\lambda)\varepsilon_B(\lambda)} \quad \text{(Eq. 3-10d)}$$

On integration, Eq. 3-10a becomes

$$\ln\left(\frac{[B]_0}{[B]_t}\right) = k_1(\lambda)t \quad \text{(Eq. 3-11)}$$

As $a_{\text{bkgd}}(\lambda)$ decreases toward 0.0, the WF(λ) tends toward 1.0. If $a_B(\lambda) < 0.02$, $\chi_B(\lambda)$ can be expanded as a Taylor series

$$\chi_B(\lambda) = 1 - 10^{-a_B(\lambda)l} \cong \ln(10)a_B(\lambda)l = \ln(10)\varepsilon_B(\lambda)l[B] \quad \text{(Eq. 3-12)}$$

This is called the *vanishing absorption* condition. On insertion into Eq. 3-7, the result is the same as Eq. 3-10a except that there WF$(\lambda) = 1.0$. Therefore, the kinetics are still first order.

However, in the situation where $a_{\text{bkgd}}(\lambda)$ decreases toward 0.0 and $a_B(\lambda) \gg 0.02$, the kinetics tend to zero order and the general rate equation (Eq. 3-7) must be used.

As an example, Figure 3-3 shows the decay of *N*-nitrosodimethylamine (NDMA) by UV photolysis. At high concentration (~1 mM), virtually all the incident UV light is absorbed, that is, $\chi_B(\lambda)$ is very close to unity. The curve scaled on the left is a straight line indicative of zero-order kinetics. When the concentration drops below ~0.01 mM, $a_B(\lambda)$ becomes small enough for the vanishing absorption approximation to apply. Now the curve scaled on the right is a straight line indicative of first-order kinetics.

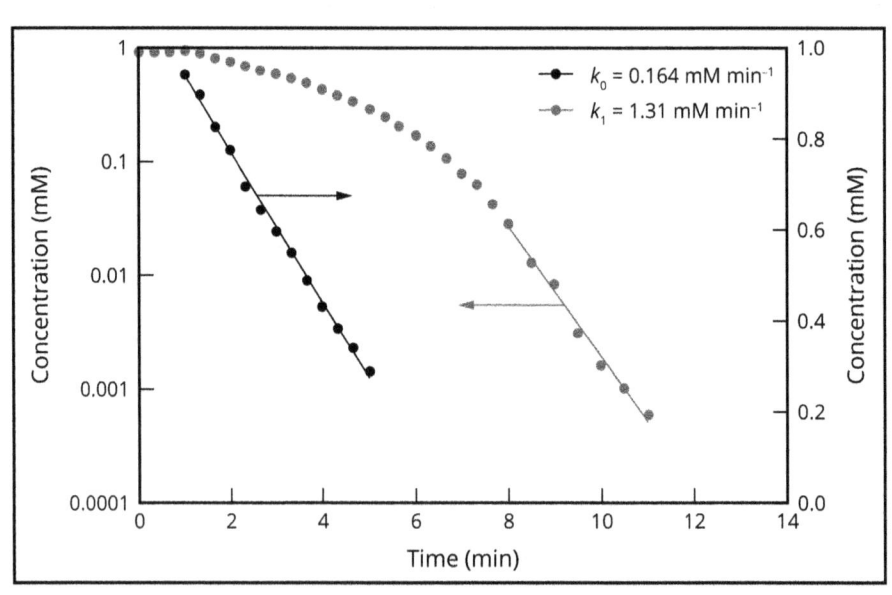

Figure 3-3 Photolysis decay of *N*-nitrosodimethylamine (NDMA) in aqueous solution. The scale on the left is a logarithmic scale and that on the right is a linear scale (adapted from Stefan and Bolton 2002).

Fluence-Based and Rate Constants

Bolton and Stefan (2002) showed that Eq. 3-11 can also be written as

$$\ln\left(\frac{[B]_0}{[B]_{F_0}}\right) = k_1'(\lambda) F_0(\text{water}) \qquad \text{(Eq. 3-13a)}$$

where $F_0(\text{water})$ is the fluence (J m^{-2}) delivered to the water. k_1' (m^2 J^{-1}) is the fluence-based rate constant given by

$$k_1'(\lambda) = \frac{\ln(10)\Phi_B(\lambda)\varepsilon_B(\lambda)}{10 U(\lambda)} \qquad \text{(Eq. 3-13b)}$$

Eq. 3-13b can be rearranged to give the quantum yield

$$\Phi_B(\lambda) = \frac{10 k_1'(\lambda) U(\lambda)}{\ln(10)\varepsilon_B(\lambda)} \qquad \text{(Eq. 3-13c)}$$

Eqs. 3-13a and 3-13b can be rearranged to derive the fluence $F_0(\text{water})_{10}$ required for 1-log (90 percent) degradation

$$F_0(\text{water})_{10} = \frac{\ln(10)}{k_1'(\lambda)} = \frac{10U(\lambda)}{\Phi_B(\lambda)\varepsilon_B(\lambda)} \qquad \text{(Eq. 3-14)}$$

Note that the fluence-based rate constant $k_1'(\lambda)$ and the fluence required for 1-log degradation $F_0(\text{water})_{10}$ are independent of any experimental parameters. Note also that the concept of a fluence-based rate constant is only applicable when the decay kinetics of B are first order.

Polychromatic Light Sources

When considering how to analyze photochemical processes driven by polychromatic light emitted over a broad range of wavelengths (e.g., from a medium-pressure UV lamp), the analysis must take account of the second law of photochemistry, namely that the effects of photons absorbed at different wavelengths must be independent and additive. Basically, the expressions developed for a monochromatic light source have to be integrated over the wavelength range in which absorption is significant. The expressions are quite complex and can be found in Bolton et al. (2015).

When using a polychromatic light source with the UV/H_2O_2 AOT, it is important to be able to determine the fraction of light absorbed by H_2O_2 over the broad spectrum of emitted light. Appendix C illustrates how this should be done for a specific example.

AOT EXPERIMENTS IN A STIRRED TANK OR MERRY-GO-ROUND REACTOR

A stirred tank reactor is one in which one or more UV lamps are placed into a stirred tank with quartz sleeves around each UV lamp. A merry-go-round reactor is one with a central UV lamp (in air) and a series of quartz tubes surrounding the lamp at a fixed distance. The quartz tubes are rotated, so that each tube receives the same UV exposure. These reactor types are convenient when larger volumes of solution have to be subjected to UV exposure. Canonica et al. (2008) carried out a study of the phototransformation of pharmaceuticals in a merry-go-round UV reactor using the previously outlined approach.

The same kinetic equations apply to these reactors as for the collimated beam apparatus, except that there is no petri factor. Also, the effective path length is not well defined. In cases like this, a photolysis standard with a known quantum yield should be used. In this way, the average fluence rate in the reactor can be determined. For example, Canonica et al. (2008) used atrazine photolysis as the standard photolysis reaction in their studies.

SUMMARY REGARDING SENSITIZED PHOTOCHEMICAL REACTIONS

Most AOT systems rely on the generation of ·OH radicals for contaminant destruction. ·OH radicals can be generated by the photolysis of an oxidant, such as H_2O_2 or O_3. Thus, contaminant destruction is achieved through indirect photolysis reactions because the degradation of the primary contaminant does not usually involve the photolysis of that contaminant. As previously indicated, contaminant destruction follows a complex sequence of oxidative degradation reactions involving the decay of the primary contaminant and its by-products.

Eq. 3-5c is the general rate expression for a sensitized photochemical reaction. It involves the kinetics of the decay of the contaminant C, which can be first order or tend to zero order, depending on the importance of scavengers. One must also consider the kinetics of the decay of B, the direct photolysis process, which can be first order in B or tend to zero order if the absorbance of B is high with a low background absorbance.

In the case of the UV/H_2O_2 AOT, often the conditions are such that the H_2O_2 absorbance is dominant, so that R_B in Eq. 3-5c is constant (zero-order kinetics). However, scavenging processes often dominate over the reaction of ·OH radicals with C, so that the decay kinetics of C are first order.

There are two fundamental competitive loss processes in light-driven AOTs. First, there is a competition for photon absorption between the activator (B) and the background absorbance. This leads to a decrease in the water factor WF(λ) in Eq. 3-10a. Second, there is a competition for ·OH radicals reacting with the target contaminant (C) versus reacting with scavengers. This leads to a decrease in the scavenging factor (SF) in Eq. 3-5c.

AOT FIGURES-OF-MERIT

Most AOTs involve a considerable input of electrical energy, such that the cost of electricity becomes a major factor in the overall treatment costs. Thus, it is understandable that figures-of-merit have been developed that are based on the efficiency in the use of electrical energy in driving the degradation processes.

Concept of Electrical Energy Dose

Most AOTs are driven by processes (e.g., UV lamps) that consume electrical energy. The *electrical energy dose* (EED) is defined as the electrical energy (kWh) consumed per unit volume (e.g., 1 m³ or 1,000 gal) of water treated. The EED may be calculated from

$$\text{EED} = \frac{1{,}000 Pt}{60 V} \qquad \text{(Eq. 3-15)}$$

where P is the electrical power (kW), t is the time (min), and V is the volume (L) of water treated. The units of EED in Eq. 3-15 are kWh/m³. If the reference volume used is 1,000 US gal, the 1,000 in the numerator should be replaced by 3785.4 to account for the conversion of liters (L) to gallons.

Electrical Energy Figures-of-Merit

The Photochemistry Commission of International Union of Pure and Applied Chemistry (IUPAC) has recommended the use of two figures-of-merit (Bolton et al. 2001), called the *electrical energy per order* (E_{EO}) and the *electrical energy per mass* (E_{EM}). The

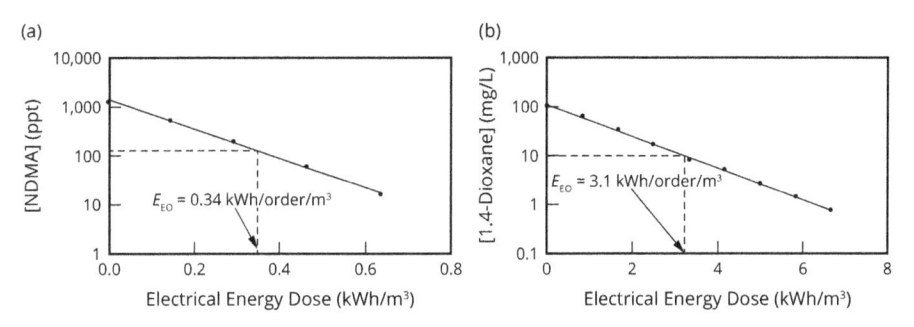

Figure 3-4 Plots of log(c) versus E_{EO}: (a) photolysis of NDMA and (b) UV/H$_2$O$_2$ treatment of 1,4-dioxane.

E_{EO} figure-of-merit applies when the concentration of the pollutant is below about 100 mg/L (usually the case) and hence the decay follows first-order kinetics. The E_{EM} figure-of-merit applies when the concentration of the pollutant is high (>100 mg/L) such that the kinetics follow the zero-order model.

The E_{EO} is defined as the electrical energy in kilowatt hours (kWh) required to bring about the degradation of a contaminant C by one order of magnitude in 1 m³ (1,000 L) of contaminated water or air. The E_{EO} can be calculated from the following expressions:

$$E_{EO} = \frac{EED}{\log(c_i/c_f)} \qquad \text{(Eq. 3-16a)}$$

$$E_{EO} = \frac{P}{\dot{V} \log(c_i/c_f)} \qquad \text{(Eq. 3-16b)}$$

where EED is the electrical energy dose (Eq. 3-15), P is the electrical power (kW), and c_i and c_f are the initial and final concentrations of the contaminant C. Eq. 3-16a is used for a batch reactor, and Eq. 3-16b is used for a flow-through reactor. $\dot{V}$ is the volume flow rate (m³/hr). By rearranging Eq. 3-16a

$$\log(c_f) = \log(c_i) - \frac{1}{E_{EO}} EED \qquad \text{(Eq. 3-17)}$$

Thus, if $\log(c_i/c_f)$ is plotted versus the EED, the E_{EO} is obtained from the negative inverse of the slope.

Note that the E_{EO} becomes smaller as the efficiency of the process increases, which results in higher levels of contaminant destruction with less energy input.

Fig. 3-4a and Fig. 3-4b show examples of how the E_{EO} can be obtained in this way.

The E_{EM} is defined as the electrical energy in kilowatt hours (kWh) required to bring about the degradation of a kilogram of contaminant C in a contaminated water or air. The E_{EM} can be calculated from the expression

$$E_{EM} = \frac{EED}{M(c_i - c_f)} \quad \text{(Eq. 3-18)}$$

where M is the molecular weight (g mol^{-1}) of C and c_i and c_f are the initial and final concentrations of C. Note that Eq. 3-18 applies only when the concentration of C is high enough so that the kinetics are zero order.

Relation of E_{EO} to Fundamental Parameters

Consider the kinetic scheme in Eq. 3-14. By analyzing the kinetic expressions, Bolton et al. (2001) showed that the E_{EO} can be expressed as

$$E_{EO} = \frac{2.422 P \sum_i k_{S_i} [S_i]}{q_p(\lambda) \chi(\lambda) \Phi_{OH} k_C} \quad \text{(Eq. 3-19)}$$

where the parameters are defined after Eq. 3-5 and Eq. 3-15. Eq. 3-19 shows that the E_{EO} is inversely proportional to Φ_{OH} and to k_C.

UV AOTs have E_{EO} values typically in the range 1–20 kWh/order/m³. The E_{EO} will be a function of site-specific factors such as hydroxyl radical scavengers, water absorbance (i.e., UVT), oxidant dose, and target contaminant concentration. As concentrations of hydroxyl radical scavengers increase (e.g., natural organic matter), so does the E_{EO} because of increased competition for ·OH radicals (see Eq. 3-5). As the overall absorbance in the 200–300 nm region increases, the E_{EO} also increases because of increased competition for photons from absorbing species other than the primary absorber (e.g., H_2O_2). For pollutants with small ·OH radical rate constants (k_C) (e.g., chloroform), the E_{EO} values are large, indicating higher energy inputs required for equivalent levels of treatment. Table 3-1 summarizes typical E_{EO} values for some groups of pollutants. As the oxidant dose increases, the E_{EO} initially decreases sharply because more photons are

Table 3-1 Typical electrical energy per order (E_{EO}) values for some groups of pollutants

Pollutant	·OH Radical Rate Constant (10⁹ M⁻¹ s⁻¹)	Electrical Energy per Order (E_{EO}) (kWh order⁻¹ m⁻³)
Benzene and its derivatives	4–7	2–10
Chlorinated alkenes (e.g., TCE)	4–7	4–20
Trichloroethylene	12	2–3
Dichloromethane	0.058	20–100
Chloroform	0.005	40–250
Carbon tetrachloride	<0.001	>250

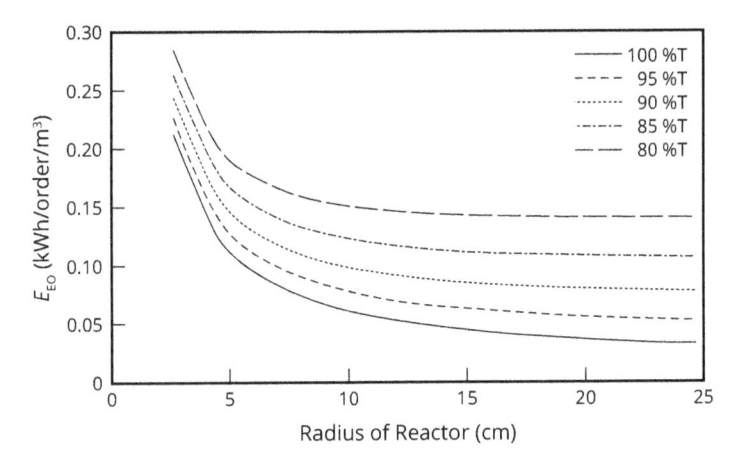

Figure 3-5 Calculated E_{EO} for an annular UV reactor (one UV lamp) as a function of the radius of the reactor and the percent transmittance of the water (adapted from Bolton and Stefan 2002).

absorbed; however, if the oxidant dose is too high, it can act as a scavenger and cause the E_{EO} to increase (see the discussion about the treatment of MTBE below).

Bolton and Stefan (2002) have shown that the E_{EO} is related to fluence-based rate constants (see above). The E_{EO} can be expressed as

$$E_{EO} = \frac{0.6396 P}{V k_1' E_o(\text{avg})} \quad \text{(Eq. 3-20a)}$$

where P is the electrical power of the UV lamps(s), V is the volume (L) of the reactor, k_1' is the fluence-based rate constant (m² J⁻¹), and $E_o(\text{avg})$ is the average fluence rate (W m⁻²) in the reactor. By inserting Eq. 3-8b, this equation becomes

$$E_{EO} = \frac{6.396 P U(\lambda)}{V \Phi_{OH} \varepsilon_C(\lambda) \ln(10) E_o(\text{avg})} \quad \text{(Eq. 3-20b)}$$

If $E_o(\text{avg})$ can be estimated (e.g., using a fluence rate distribution model), Eq. 3-20b can be used to estimate the E_{EO} for a given UV reactor. Eq. 3-20b shows that the E_{EO} is inversely proportional to the quantum yield and the molar absorption coefficient of the contaminant C.

Note that the E_{EO} can vary according to the size of the reactor. Figure 3-5 shows the E_{EO} as a function of the radius of an annular UV reactor for various conditions (Bolton and Stefan 2002). Generally, the calculated E_{EO} decreases (higher efficiency) as the radius increases. The reason for this is that for the larger radii, more of the UV is absorbed in the solution, rather than being absorbed at the walls. Note that as the UVT decreases (i.e.,

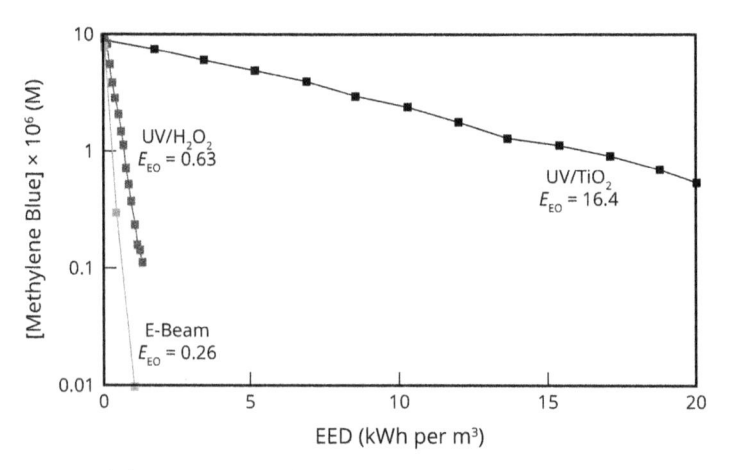

Figure 3-6 Log of the concentration of methylene blue versus EED (adapted from Bolton et al. 1998).

increased UV absorbance), the curve reaches a constant plateau. As UVT decreases, the UV is absorbed more rapidly, and so beyond a certain distance, there is no more UV to react with the target contaminant or oxidant.

Comparison of AOT Efficiencies

Bolton et al. (1998) evaluated the relative treatment efficiencies of three AOTs (UV/H_2O_2, electron beam, and UV/TiO_2) for the degradation of methylene blue. Figure 3-6 shows the results.

The UV/H_2O_2 and the electron beam processes comparable E_{EO}s; however, the UV/TiO_2 process is much higher. This undoubtedly arises from the very low hydroxyl radical quantum yield (~0.04) in that case (Sun and Bolton 1996).

Scale-up to Full-Scale Reactors

In full-scale UV reactors, generally, the effective path lengths are much larger than in the case of a bench-scale UV reactor. In reference to Fig. 3-5, it is clear that the full-scale UV reactor will have a lower E_{EO} than the bench-scale reactor. If this feature is not understood, reactor designs based on bench-scale studies may be considerably overdesigned.

The Dose per Log Concept

Bircher et al. (2012) introduced the *dose per log* (D_L) concept, which they found useful in the scale-up of bench-scale results to full-scale UV reactors. Dose calculations for AOT operation would differ from disinfection applications. In an AOT application, the UV dose would be calculated based on the absorbance of the oxidant or target contaminant and not the absorption of DNA. From Eq. 3-14, it is seen that the fluence for 1 log decrease (same as dose per log) in the activator concentration (B) is independent of any

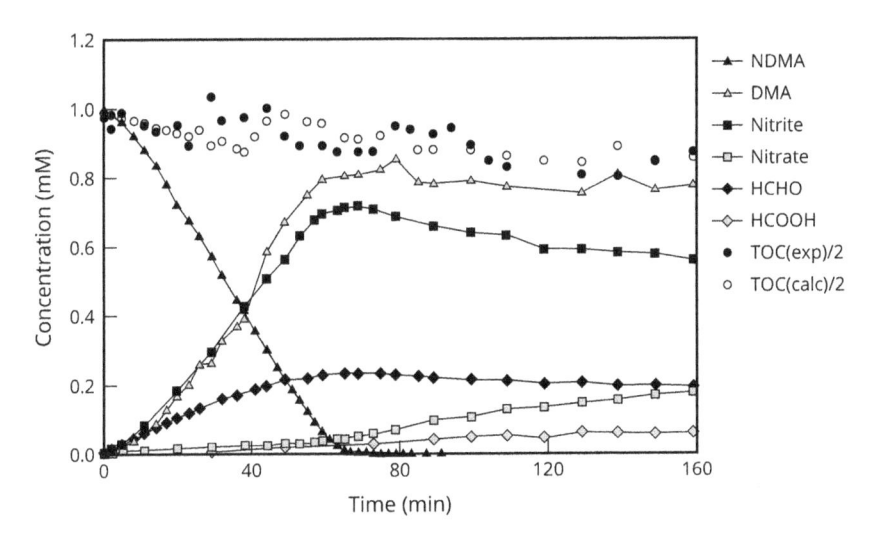

Figure 3-7 Photolysis decay of NDMA and the growth of several products. TOC(exp) is the experimental total organic carbon (TOC) and TOC(calc) is the TOC calculated by adding the carbon content of all identified components (adapted from Stefan and Bolton 2001).

experimental parameters. Because the primary photolysis rate also appears in the expression for the rate of a sensitized photochemical reaction (Eq. 3-5c), the D_L concept also applies to sensitized photochemical reactions. Under this concept, the UV dose required for contaminant reduction at the bench-scale would be equal to the UV dose required at the full-scale. As with E_{EO}, the D_L would be site-specific and a function of the hydroxyl radical scavengers, target contaminant, and oxidant dose. The D_L concept would account for the water UV absorbance or UVT as the dose calculations take into account the UV absorbance of the water.

AOT EXAMPLES

The following sections provide examples of various AOTs for a range of potential contaminants.

Photodegradation of NDMA

UV AOTs are very effective in treating *N*-nitrosodimethylamine (NDMA). NDMA undergoes direct photolysis at wavelengths <260 nm. UV/H_2O_2 treatment, which involves generation of ·OH radicals, can also be used, but high concentration ratios of H_2O_2 to NDMA are required. The direct photolysis of NDMA has been studied extensively and a full mechanism of the photodegradation has been established (Stefan and Bolton 2002).

Direct photolysis AOTs have been used extensively in controlling NDMA in contaminated groundwaters and in many potable reuse applications.

Figure 3-7 shows the photolysis by-products, which are principally dimethylamine and the nitrite ion.

UV/H$_2$O$_2$ Treatment of MTBE

Methyl-*t*-butyl ether (MTBE) is a fuel oxygenate used as an octane enhancer of reformulated gasoline and is used extensively in the United States. It has a high solubility in water and has been detected in groundwater and storm water as the second most frequent contaminant [after chloroform (CHCl$_3$)]. MTBE is known to be a carcinogen in animals and a potential human carcinogen. It is not currently regulated as a drinking water contaminant. However, the USEPA has a drinking water advisory limit of 20–40 µg/L for MTBE. Detection thresholds for MTBE are relatively low with an odor threshold of 45 ppb and a taste threshold of 39 ppb.

Traditional technologies have not proven effective in treating MTBE.

- Air-stripping—can achieve 99 percent removal of MTBE from water if large air-to-water ratios are used but is only a mass transfer process.

- Adsorption on granulated activated carbon—low affinity; effective at low concentrations, but a high cost of carbon replacement at high concentrations.

- Aerobic biodegradation—difficult to apply to large volumes of MTBE-contaminated water or to ppm-ppb levels.

AOTs have been proven as an effective treatment technique. Cater et al. (2000) evaluated UV/H$_2$O$_2$ for the treatment of MTBE. They assumed a steady-state kinetic model and assumed that the rate of MTBE degradation can be estimated from initial conditions. The rate of removal of MTBE is

$$\text{Rate} = k_2 [\cdot OH]_{SS} [\text{MTBE}] \qquad \text{(Eq. 3-20)}$$

The steady-state ·OH concentration is given by (see Eq. 3-5b)

$$[\cdot OH]_{SS} = \frac{q_p \chi_{H_2O_2} \Phi_{H_2O_2}/V}{k_{\text{MTBE}}[\text{MTBE}] + k_{H_2O_2}[H_2O_2] + \sum_i k_{S_i}[S_i]} \qquad \text{(Eq. 3-21)}$$

where q_p is the incident photon flux (einstein sec^{-1}), $\chi_{H_2O_2}$ is the fraction of the incident photon flux absorbed by H$_2$O$_2$; $\Phi_{H_2O_2}$ is the quantum yield for generation of ·OH radicals; k_{MTBE} is the rate constant for reaction of ·OH radicals with MTBE; $k_{H_2O_2}$ is the rate constant for reaction of ·OH radicals with H$_2$O$_2$; and k_{S_i} is the rate constant for reaction of ·OH radicals with various scavengers S_i.

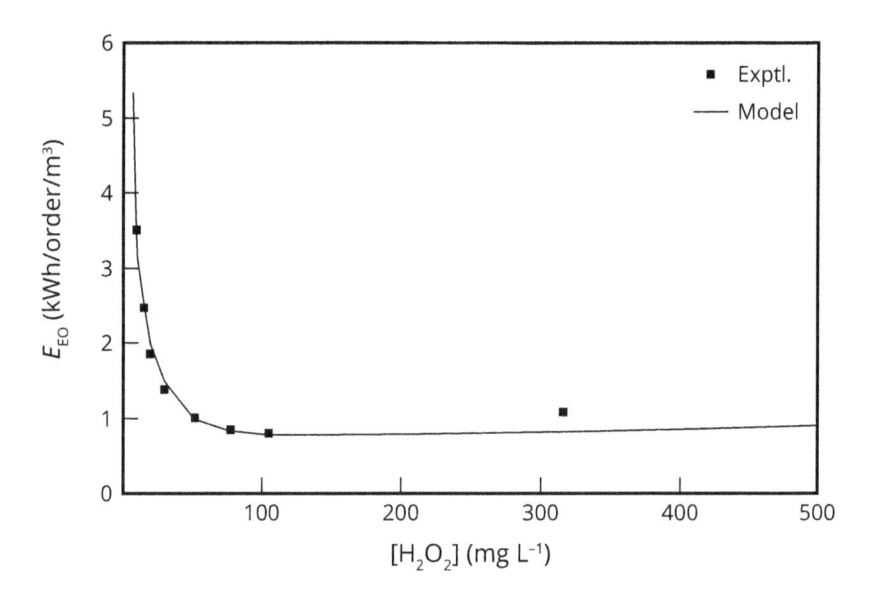

Figure 3-8 E_{EO} **as a function of the H$_2$O$_2$ concentration in the UV/H$_2$O$_2$ degradation of MTBE (adapted from Cater et al. 2000).**

Figure 3-8 shows the influence of H$_2$O$_2$ on the E_{EO} (Cater et al. 2000). At low concentrations, the fraction of UV light absorbed by H$_2$O$_2$ is low. A sharp decrease in the E_{EO} arises from an increased fraction of light absorbed by H$_2$O$_2$. At high concentrations, most of the light is absorbed by H$_2$O$_2$, but the H$_2$O$_2$ acts as an added scavenger (see Eq. 3-21) resulting in the E_{EO} gradually increasing. The solid line in Figure 3-8 shows that the kinetic model works quite well.

Stefan et al. (2000) have studied the mechanism of the degradation of MTBE by the UV/H$_2$O$_2$ process. Figures 3-9a, b, and c show the progression of intermediates.

Figure 3-10 shows the time progression of the concentration of H$_2$O$_2$ and the calculated and measured TOC. The calculated TOC was obtained by adding the carbon content of MTBE and all detected intermediates at that time. The fact that the calculated and measured TOC values agree indicates that most of the intermediates have been detected.

Note that the H$_2$O$_2$ decays according to zero-order kinetics. This example represents one of the most detailed studies of the treatment of a pollutant by the UV/H$_2$O$_2$ process. The process was carried out to the stage of almost complete mineralization, as indicated by the fact that the TOC decreased almost to zero.

Finally, it should be noted that the UV/H$_2$O$_2$ AOT is relatively insensitive to pH, except where the alkalinity is high. In that case, the reaction rate slows down at higher pH because carbonate (CO_3^{2-}) is a much stronger scavenger than is bicarbonate (HCO_3^{-}).

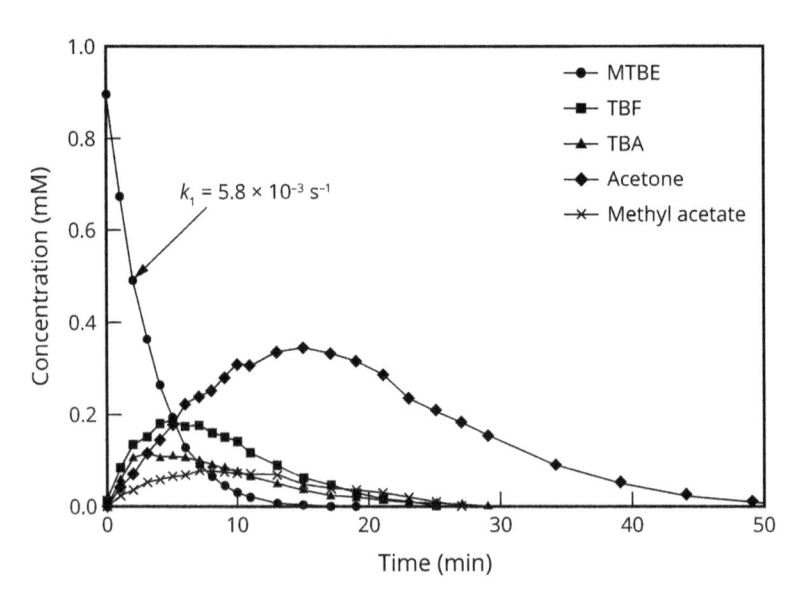

Figure 3-9a Concentration of MTBE and some of the intermediates as a function of time. TBF is *t*-butyl formate; TBA is *t*-butyl alcohol (adapted from Stefan et al. 2000).

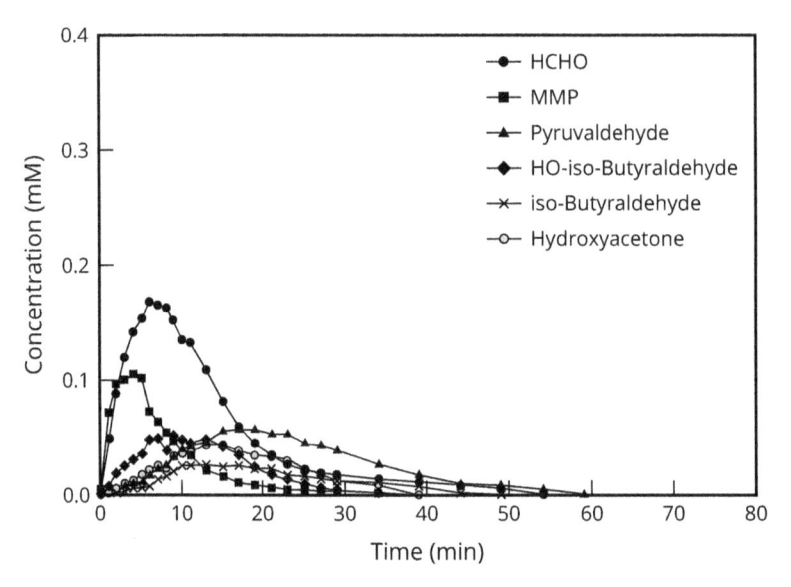

Figure 3-9b Concentration of selected intermediates as a function of time in the degradation of MTBE. MMP is 2-methoxy-2-methyl propionaldehyde (adapted from Stefan et al. 2000).

Fundamentals of Advanced Oxidation

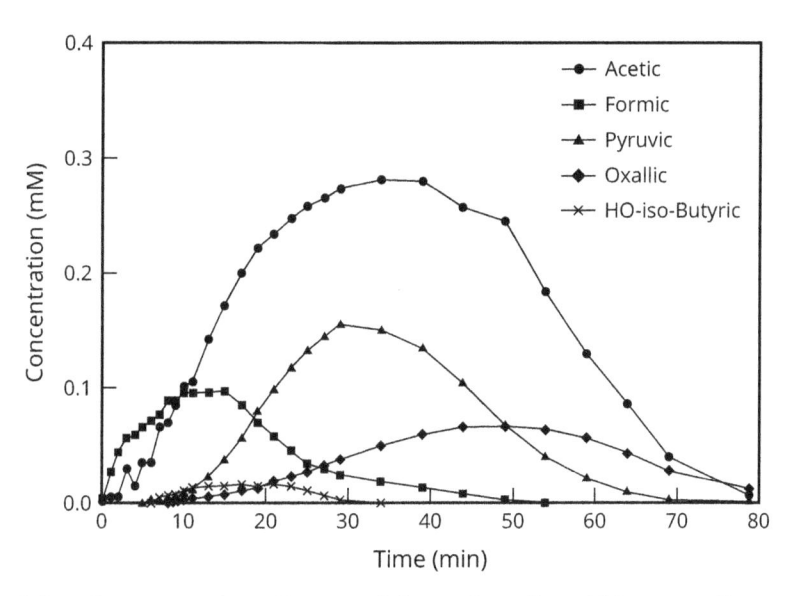

Figure 3-9c Concentration of some of the carboxylic acid intermediates as a function of time in the degradation of MTBE. The intermediates shown here are all carboxylic acids (adapted from Stefan et al. 2000).

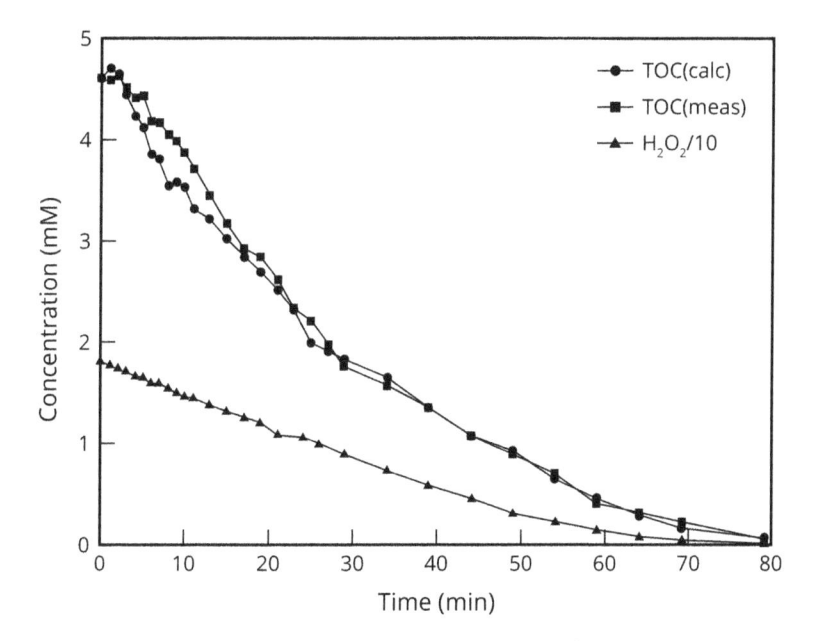

Figure 3-10 Concentration of H_2O_2 and the calculated and measured TOC for the UV/H_2O_2 degradation of MTBE as a function of time (adapted from Stefan et al. 2000).

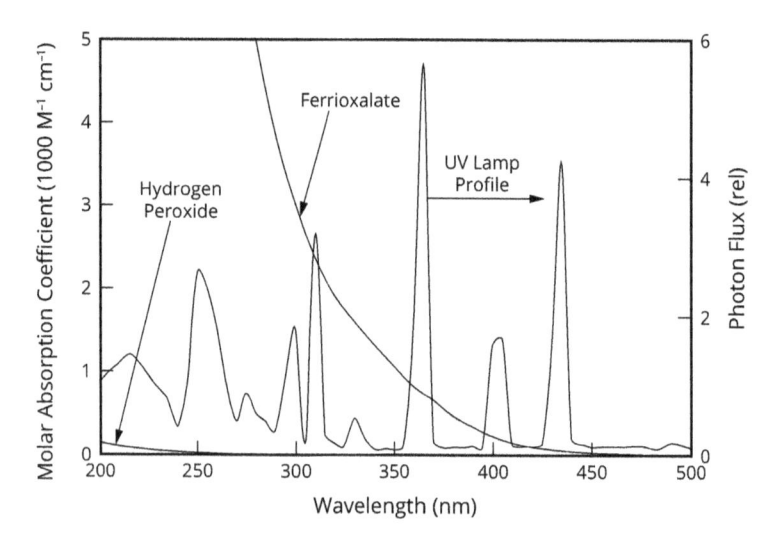

Figure 3-11 Comparison of the absorption spectra of ferrioxalate and hydrogen peroxide with the emission spectrum of a medium-pressure UV lamp (courtesy Calgon Carbon Corp.).

Photo-Fenton's AOT

The UV-Fenton's AOT using ferrioxalate and H_2O_2 has been employed effectively to destroy waters contaminated with BTEX (benzene, ethylbenzene, toluene, and xylenes) from leaking underground storage tanks. This process makes efficient use of a medium-pressure lamp output because of the absorption of ferrioxalate over the UV and visible range, as seen in Figure 3-11.

Recall the photochemistry of ferrioxalate

$$Fe^{III}(C_2O_4)_3^{3-} + h\nu \rightarrow Fe^{2+} + 2.5\ C_2O_4^{2-} + CO_2 \qquad \text{(Eq. 3-22a)}$$

$$Fe^{2+} + H_2O_2 \rightarrow Fe^{3+} + \cdot OH + OH^- \qquad \text{(Eq. 3-22b)}$$

The ferrous ion has a high reactivity with hydrogen peroxide to yield ·OH radicals. The high quantum yield of Fe(II) formation (1.39) means a very high rate of generation of hydroxyl radicals. Also photolysis of Fe(III)-organic intermediate complexes enhances the treatment effectiveness.

Figure 3-12 illustrates a comparison of the UV-vis/ferrioxalate/H_2O_2 photo-Fenton's process with the UV/H_2O_2 process for the treatment of a water containing 15 mg/L of BTEX. Note that the UV-vis/ferrioxalate/H_2O_2 process requires that the pH be reduced to 3.

Fundamentals of Advanced Oxidation

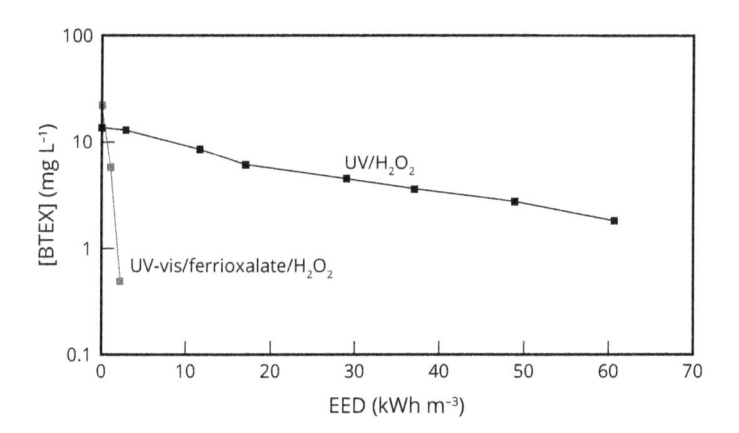

Figure 3-12 Comparison of the UV-vis/ferrioxalate/H$_2$O$_2$ process and the UV/H$_2$O$_2$ process for the treatment of a water contaminated with 15 mg/L of BTEX (courtesy Calgon Carbon Corp.).

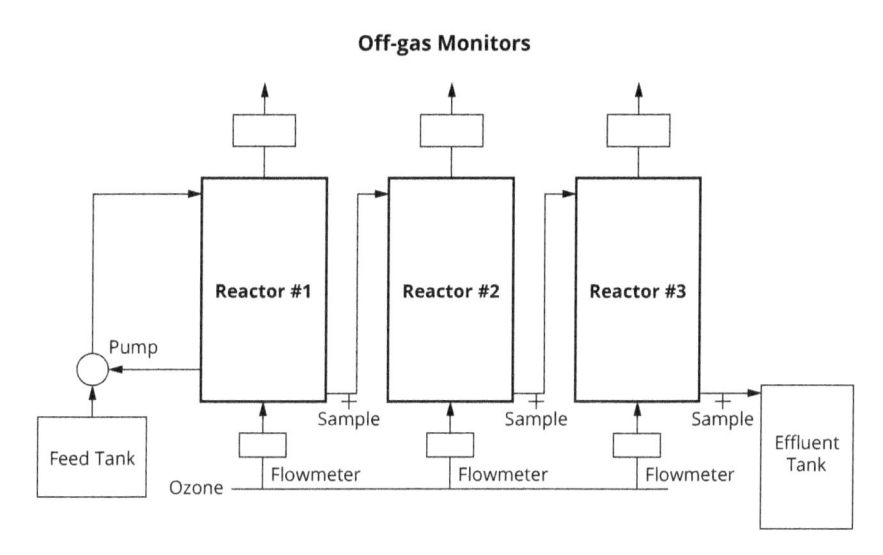

Figure 3-13 Experimental setup for the treatment of MTBE by the O$_3$/H$_2$O$_2$ AOT (from Safarzadeh-Amiri 2001).

O$_3$/H$_2$O$_2$ Treatment of MTBE

Safarzadeh-Amiri (2001) described a bench-scale study of the degradation of MTBE by the O$_3$/H$_2$O$_2$ process using a ratio of O$_3$ to H$_2$O$_2$ of 0.33 (close to the stoichiometric value of 0.354). The study evaluated MTBE concentrations from 0.05 to 80 mg/L and H$_2$O$_2$ concentrations from 150 to 350 mg/L. Figure 3-13 shows the apparatus that was used.

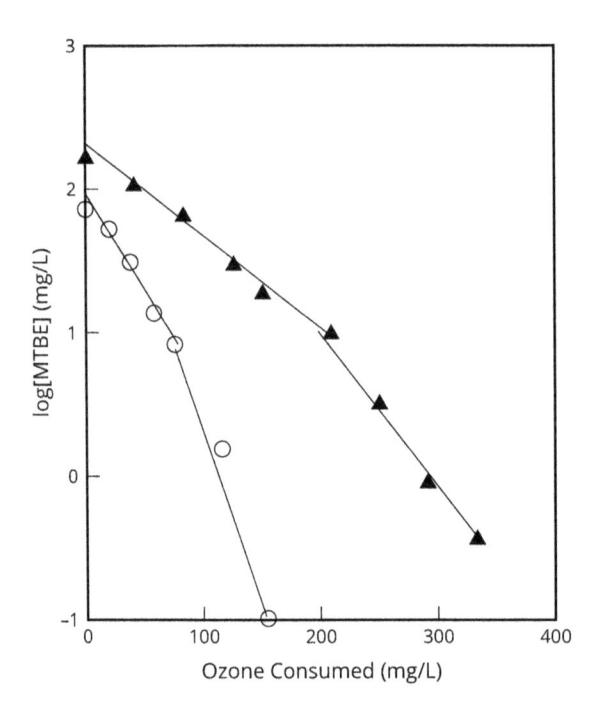

Figure 3-14 Treatment of MTBE-spiked tap water and an MTBE-contaminated groundwater at the bench scale with the O$_3$/H$_2$O$_2$ process (from Safarzadeh-Amiri 2001).

Figure 3-14 shows the decay of MTBE as a function of the ozone dose for tap water spiked with MTBE and for a groundwater contaminated with MTBE.

In this case, it was determined that the electrical energy consumption to produce ozone is 22 kWh/kg.[4] There are clearly two kinetic phases: a *slow phase*, where the E_{EO} is 1.7 kWh/order/m^3, and a *fast phase*, where the E_{EO} is 0.9 kWh/order/m^3, these values being for the MTBE-spiked tap water. This compares with 0.8 kWh/order/m^3 for the UV/H$_2$O$_2$ process for [H$_2$O$_2$] = 100 mg/L (Cater et al. 2000).

Liang et al. (2001) also studied the treatment of MTBE with the peroxone process at the pilot scale using the apparatus shown in Fig. 3-15.

Figures 3-16 a and b shows the treatment results for two different MTBE concentrations.

The applied ozone dose for 90 percent removal of MTBE for the O$_3$/H$_2$O$_2$ process was about 6 mg/L, or 6 g/m^3, or 0.006 kg/m^3. Each kilogram of O$_3$ requires 22.5 kWh.

---

[4] Dr. Amiri has since pointed out in a private communication that the 22 kWh/kg only applies for ozone generated from pure oxygen and does not include the energy required to produce pure oxygen. When these energy terms are added in, the figure rises to about 35–40 kWh/kg, and thus the E_{EO}s will be correspondingly smaller.

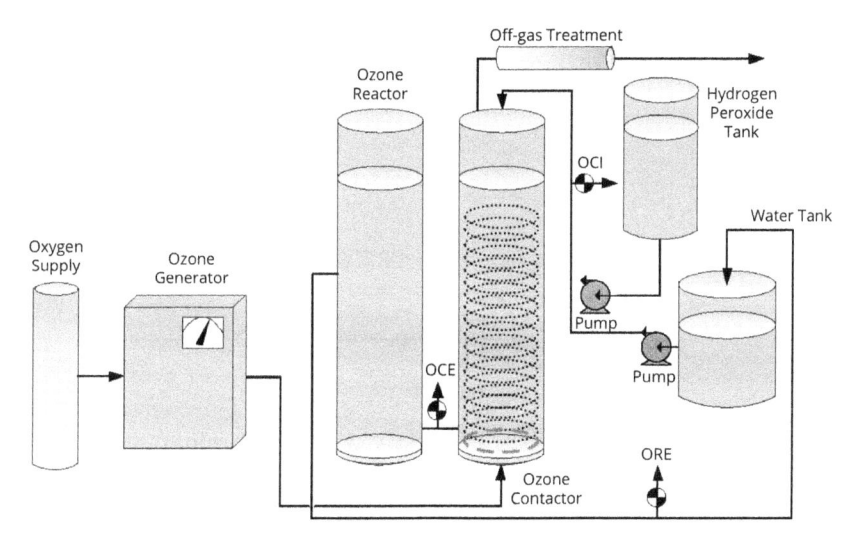

OCI—ozone contactor influent, OCE—ozone contactor effluent, ORE—ozone reactor effluent

Figure 3-15 Apparatus for the treatment of an MTBE-contaminated groundwater at the pilot scale with the O$_3$ and O$_3$/H$_2$O$_2$ processes (from Liang et al. 2001).

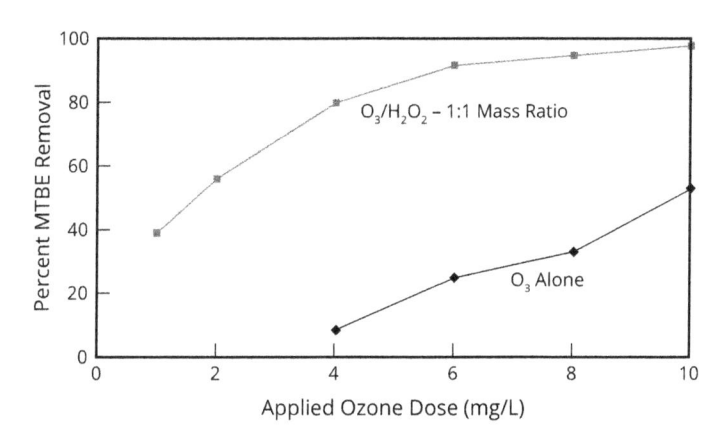

Figure 3-16a Percent removal of MTBE in an MTBE-contaminated groundwater using the O$_3$/H$_2$O$_2$ process and O$_3$ alone for an MTBE concentration of 2.23 mg/L (from Liang et al. 2001).

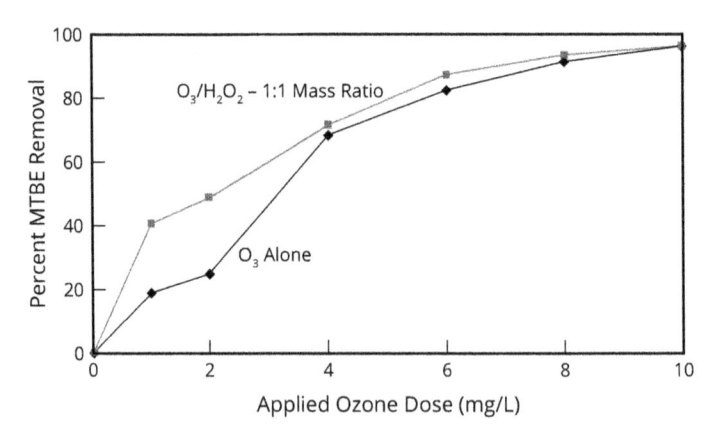

Figure 3-16b Percent removal of MTBE in a MTBE-contaminated groundwater using the O_3/H_2O_2 process and O_3 alone for an MTBE concentration of 0.18 mg/L (from Liang et al. 2001)

Therefore, the E_{EO} is $0.006 \times 22.5 = 0.135$ kWh/order/m^3. This is smaller than the value of 0.9 kWh/order/m^3 obtained by Safarzadeh-Amiri (2001) at the bench scale.

Rosenfeldt et al. (2006, 2008) have compared the UV/H_2O_2 (LP and MP) and the O_3 and O_3/H_2O_2 AOTs at the bench scale in the treatment of p-chlorobenzoic acid at the bench scale. They found that the ozone AOTs were much more energy efficient than the UV AOTs, with the LP UV/H_2O_2 AOT being more energy efficient than the MP UV/H_2O_2 AOT. However, they noted in an erratum (Rosenfeldt et al. 2008) that the path length for the UV/H_2O_2 AOT was only 0.95 cm. The longer path length that would prevail in a full-scale UV/H_2O_2 UV reactor would make the comparison of energy efficiency with the ozone-based AOTs more comparable.

REFERENCES

Bircher, K., M. Vuong, B. Crawford, M. Heath, and J. Bandy. 2012. Using UV Dose Response for Scale Up of UV/AOP Reactors. In *Proc. of the International Water Association World Congress*, Busan, Korea.

Bolton, J.R. 2010. *Ultraviolet Applications Handbook*, 3rd Ed. Edmonton, AB, Canada: ICC Lifelong Learn Inc.

Bolton, J.R., and K.G. Linden. 2003. Standardization of Methods for Fluence (UV Dose) Determination in Bench-Scale UV Experiments. *Jour. Environ. Eng.*, 129(3):209–216.

Bolton, J.R., and M.I. Stefan. 2002. Fundamental Photochemical Approach to the Concepts of Fluence (UV Dose) and Electrical Energy Efficiency in Photochemical Degradation Reactions, *Res. Chem. Intermed.* 28(7-9):857–870.

Bolton, J.R., J.E. Valladares, J.P. Zanin, W.J. Cooper, M.G. Nickelsen, D.C. Kajdi, T.D. Waite, and C.N. Kurucz. 1998. Figures-of-Merit for Advanced Oxidation Technologies: A Comparison of Homogeneous UV/H_2O_2, Heterogeneous TiO$_2$ and Electron Beam Processes, *Jour. Advan. Oxid. Technol.*, 3:174–181.

Bolton, J.R., K.G. Bircher, W. Tumas, and C.A. Tolman. 2001. Figures-of-Merit for the Technical Development and Application of Advanced Oxidation Technologies for Both Electric- and Solar-Driven Systems. *Pure Appl. Chem.,* 73(4):627–637.

Bolton, J.R., I. Major-Smith, and K.G. Linden. 2015. Rethinking the Concepts of Fluence (UV Dose) and Fluence Rate: The Importance of Photon-Based Units—A Systemic Review. *Photochem. Photobiol.*

Calgon Carbon Corp. 2015. Data kindly provided.

Canonica, S., L. Meunier, and U. von Gunten. 2008. Phototransformation of Selected Pharmaceuticals During UV Treatment of Drinking Water. *Wat. Res.,* 42:121–128.

Cater, S.R., M.I. Stefan, J.R. Bolton, and A. Safarzadeh-Amiri. 2000. Degradation Pathways During the Treatment of Methyl-*tert*-butyl Ether by the UV/H_2O_2 Process. *Environ. Sci. Technol.,* 34:650–658.

Daimon, M., and A. Masumura. 2007. Measurement of the Refractive Index of Distilled Water From the Near-Infrared Region to the Ultraviolet Region. *Appl. Optics.,* 46(18):3811–3820.

Leifer, A. 1988. *The Kinetics of Environmental Aquatic Photochemistry: Theory and Practice,* Section 1C.4, Washington, DC: American Chemical Society.

Liang, S., R.S. Yates, D.V. Davis, S.J. Pastor, L.S. Palencia, and J.-M. Bruno. 2001. Treatability of MTBE-Contaminated Groundwater by Ozone and Peroxone. *Jour. AWWA,* 93(6):110–120.

NDRL/NIST. 2015. NDRL/NIST Solution Kinetics Database on the Web. http://kinetics.nist.gov/solution/. Last accessed on 23 May 2015.

Rosenfeldt, E.J., K.G. Linden, A. Canonica, and U. von Gunten. 2006. Comparison of the Efficiency of ·OH Radical Formation During Ozonation and the Advanced Oxidation Processes O_3/H_2O_2 and UV/H_2O_2. *Wat. Res.,* 40:3695–3704.

Rosenfeldt, E.J., K.G. Linden, A. Canonica and U/ von Gunten. 2008. Erratum to "Comparison of the Efficiency of ·OH Radical Formation During Ozonation and the Advanced Oxidation Processes O_3/H_2O_2 and UV/H_2O_2." *Wat. Res.,* 42:2836–2838.

Safarzadeh-Amiri, A. 2001. O_3/H_2O_2 Treatment of Methyl-*tert*-butyl Ether (MTBE) in Contaminated Waters. *Wat. Res.,* 35(15):3706–3714.

Stefan, M.I., and J.R. Bolton. 2002. UV Direct Photolysis of *N*-nitrosodimethylamine (NDMA): Kinetic and Product Study. *Helv. Chim. Acta,* 85:1416–1426.

Stefan, M.I., J. Mack, and J.R. Bolton. 2000. Degradation Pathways During the Treatment of Methyl-*tert*-Butyl Ether by the UV/H_2O_2 Process. *Environ. Sci. Technol.,* 34:650–658.

Sun, L. and J.R. Bolton. 1996. Determination of the Quantum Yield for the Photochemical Generation of Hydroxyl Radicals in TiO_2 Suspensions. *Jour. Phys. Chem.,* 100:4127–4134.

4

Advanced Oxidation Types[1]

In chapters 2 and 3, the fundamentals of advanced oxidation technologies (AOTs) were explained. In this chapter, the various AOT types, most of which involve the generation of ·OH radicals by some means, are examined. Most AOTs occur in a homogenous solution, meaning they occur in one phase (e.g., liquid phase). However, some are heterogeneous processes, meaning that they occur in two phases (e.g., solid and liquid phases). This chapter presents a range of AOT technologies; however, not all AOTs are commercially viable at this time. The remaining chapters of this book focus on UV and ozone-based applications, as they are the most commonly used commercial-scale AOTs.

LIGHT-DRIVEN HOMOGENEOUS AOTS

Light-driven homogeneous AOTs are those that involve absorption of UV and/or visible light in a homogeneous aqueous solution. Photochemical processes in solution lead to the generation of ·OH radicals, which initiate the oxidation and degradation of the organic pollutants. An important exception is direct photolysis, such as the use of UV to photolyze *N*-nitrosodimethylamine (NDMA) directly in contaminated waters.

DIRECT PHOTOLYSIS

In direct photolysis, the contaminant absorbs UV light directly. Although ·OH radicals are usually not generated, these processes are generally considered to be part of the AOT family.

The molar absorption coefficients (i.e., the absorbance of a compound for a concentration of 1 M and a path length of 1 cm) must be high (>1,000 M^{-1} cm^{-1}) at wavelengths where the lamp emission is strong. Direct photolysis examples are: *N*-nitrosodimethylamine (NDMA), trichloroethylene (TCE), free chlorine (HOCl or OCl$^-$), and certain pesticides and herbicides (e.g., atrazine).

Figure 4-1 shows the absorption spectra of some pollutants compared to the emission spectrum of a medium-pressure UV lamp. In order for photolysis to occur, the contaminant of concern must absorb UV light in the range emitted by the UV lamp.

---

[1] Some of this chapter has been adapted (with permission) from Bolton (2010).

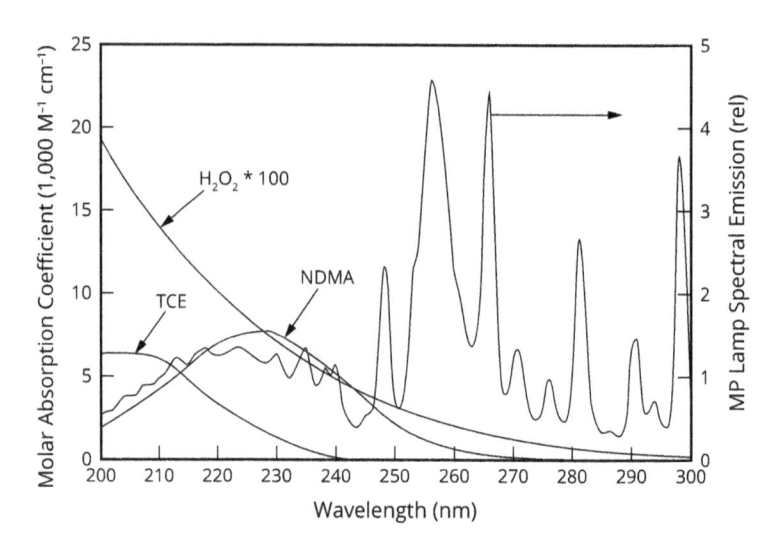

Figure 4-1 Absorption spectra of some pollutants [trichloroethylene (TCE) and *N*-nitrosodimethylamine (NDMA)] and the emission spectrum of a medium-pressure UV lamp.

The UV/O$_3$ Process

The photolysis of ozone (O$_3$) in the 200–280 nm region (UVC) can lead to the generation of ·OH radicals through the reactions (Glaze et al. 1987).

$$O_3 + h\nu \rightarrow O_2 + O(^1D) \tag{Eq. 4-1a}$$

$$O(^1D) + H_2O \rightarrow [HO\cdot \ldots \cdot OH] \rightarrow H_2O_2 \tag{Eq. 4-1b}$$

$$H_2O_2 + h\nu \rightarrow 2\cdot OH \tag{Eq. 4-1c}$$

where the square brackets in reaction Eq. 4-1b represent a solvent cage, in which almost all of the hydroxyl radical pairs combine to form H$_2$O$_2$ within the cage. Hence, to generate OH radicals, another photon must be absorbed by the generated H$_2$O$_2$.

The UV/O$_3$ process has been used commercially, particularly in the treatment of groundwaters containing contaminants such as trichloroethylene (TCE), perchloroethylene (PCE), or trinitrotoluene (TNT); however, for most applications, it is not considered economically feasible compared to the UV/H$_2$O$_2$ or O$_3$/H$_2$O$_2$ processes.

The VUV Water Photolysis Process

Water absorbs UV in the *vacuum ultraviolet* (VUV) region (<200 nm) to undergo the following photochemical reaction (Oppenländer 2003):

$$H_2O + h\nu \rightarrow H\cdot + \cdot OH \qquad \text{(Eq. 4-2)}$$

with a quantum yield for generation of ·OH radicals of 0.42 (Heit et al. 1998).

The usual light sources for this process are ozone-producing low-pressure mercury lamps (emitting at 185 nm) and the Xe excilamp (emitting at 172 nm).

The molar absorption coefficient of H_2O increases sharply at wavelengths < 190 nm, so most of the UV is absorbed within a few millimeters or less. For example, at 185 nm, the absorption depth[2] in water is 5.6 mm, whereas at 172 nm, the absorption depth in water is 0.017 mm (Weeks et al. 1963).

The advantage of this process is that no additional chemicals are required. For this reason, VUV is used in the treatment of ultrapure water in the semiconductor industry.

The UV/H$_2$O$_2$ Process

The UV/H$_2$O$_2$ is by far the most important commercial UV-based AOT. It is based on the direct photolysis of added hydrogen peroxide (Bolton and Cater 1994, Beltrán 2003, Tuhkanen 2004)

$$H_2O_2 + h\nu \rightarrow 2\cdot OH \qquad \text{(Eq. 4-3)}$$

The quantum yield for the generation of ·OH radicals is 1.11±0.07 (Goldstein et al. 2007). The molar absorption coefficients of H_2O_2 are low (see Figure 4-1). Therefore, sufficient H_2O_2 must be added (usually >5 mg/L), so that a significant fraction of the UV light between 200 and 300 nm is absorbed by the H_2O_2. UV lamps that emit strongly below 250 nm (e.g., a medium-pressure mercury lamp) may be advantageous because of the higher adsorption of H_2O_2 below 250 nm. However, if the water absorbs UV strongly in the 200–300 nm region and/or the alkalinity is very high (for pH >7), such that scavenging by bicarbonate/carbonate ions is important, this process may not be as effective.

The UV/Chlorine Process

The UV/Chlorine process is similar to the UV/H$_2$O$_2$ process except that chlorine is the added oxidant in place of H_2O_2 (Watts and Linden 2007, Feng et al. 2007). When Cl_2 dissolves in water, it reacts with water to form hypochlorous acid (HOCl).

$$Cl_2 + H_2O \rightarrow HOCl + HCl \qquad \text{(Eq. 4-4a)}$$

HOCl is a weak acid and readily dissociates to form the hypochlorite ion (OCl$^-$).

---

2 The absorption depth is defined as the depth at which the UV irradiance has dropped to one-tenth of its value at the surface.

$$\text{HOCl} \leftrightarrow \text{OCl}^- + \text{H}^+ \qquad \text{p}K_a = 7.6 \qquad \text{(Eq. 4-4b)}$$

Both HOCl and OCl⁻ absorb UV photons and the result is reactive species, including hydroxyl and chlorine radicals.

$$\text{HOCl} + h\nu \rightarrow \cdot\text{OH} + \cdot\text{Cl} \qquad \text{(Eq. 4-4c)}$$

$$\text{OCl}^- + h\nu \rightarrow \cdot\text{O}^- + \cdot\text{Cl} \qquad \text{(Eq. 4-4d)}$$

$$\cdot\text{O}^- + \text{H}_2\text{O} \rightarrow \cdot\text{OH} + \text{OH}^- \qquad \text{(Eq. 4-4e)}$$

The quantum yields for reactions shown in Eqs. 4-4c and 4-4d are near 1.0.

OCl⁻ absorbs strongly in the 300–400 nm region. Chan et al. (2012) and Shu et al. (2014) have shown that this process can be driven efficiently using sunlight.

The UV-vis/Fenton's Processes

UV-vis/Fenton's processes are based on the photoreduction of ferric ion (Fe³⁺) and ferric complexes, which result in Fe²⁺ ions capable of reacting with H₂O₂ in a Fenton's reaction (Eq. 4-5b) to generate ·OH radicals. UV-vis/Fenton's processes may be effective for contaminated waters that strongly absorb in the 200–300 nm range and when contaminant concentrations are high (Tarr 2003, Wadley and Waite 2004). These processes require a pH of about 3, so there are added costs of pH adjustment. The simplest UV-vis/Fenton's process is the photolysis of $\text{Fe}^{\text{III}}(\text{OH})^{2+}$.

$$\text{Fe}^{\text{III}}(\text{OH})^{2+} + h\nu \rightarrow \text{Fe}^{2+} + \cdot\text{OH} \qquad \text{(Eq. 4-5a)}$$

$$\text{Fe}^{2+} + \text{H}_2\text{O}_2 \rightarrow \text{Fe}^{3+} + \cdot\text{OH} + \text{OH}^- \qquad \text{(Eq. 4-5b)}$$

The quantum yield of reaction Eq. 4-5a is about 0.21 (Nadtochenko and Kiwi 1998), and the $\text{Fe}^{\text{III}}(\text{OH})^{2+}$ ion absorbs UV to 400 nm.

Fe(III) forms complexes with many organic acids. One of the most effective is the trioxalato complex with oxalic acid. The overall reactions are

$$\text{Fe}^{\text{III}}(\text{C}_2\text{O}_4)_3^{3-} + h\nu \rightarrow \text{Fe}^{2+} + 2.5\,\text{C}_2\text{O}_4^{2-} + \text{CO}_2 \qquad \text{(Eq. 4-5c)}$$

$$\text{Fe}^{2+} + \text{H}_2\text{O}_2 \rightarrow \text{Fe}^{3+} + \cdot\text{OH} + \text{OH}^- \qquad \text{(Eq. 4-5d)}$$

The quantum yield of reaction Eq. 4-5c is 1.39 ±0.02 at 254 nm (Goldstein and Rabani 2008, Bolton et al. 2011), and ferrioxalate absorbs out to 500 nm. This makes ferrioxalate a very efficient form of the UV-vis/Fenton's process. Because ferrioxalate absorbs well into the visible region, this process can be driven by solar light. However, the addition of oxalate and acid to lower the pH adds to the chemical costs.

DARK HOMOGENEOUS AOTS

Several AOTs do not involve the use of visible or UV light. These are known as *dark* AOTs.

The O$_3$/H$_2$O$_2$ (Peroxone) Process

The most important dark AOT is the O$_3$/H$_2$O$_2$ process (also known as the *peroxone process*). H$_2$O$_2$ acts as a promoter in the ozone decomposition cycle, which enhances the generation of ·OH radicals. In many cases, the O$_3$/H$_2$O$_2$ process has similar economics to the UV/H$_2$O$_2$ process.

The O$_3$/H$_2$O$_2$ (peroxone) process has been applied at full scale in several applications. H$_2$O$_2$ acts to promote the ozone decomposition cycle, shifting the process toward a greater production of ·OH radicals. Hydroxyl radical generation proceeds through the following reactions:

$$H_2O_2 \leftrightarrow HO_2^- + H^+ \quad\quad\quad (Eq.\ 4\text{-}6a)$$

$$O_3 + HO_2^- \rightarrow \cdot OH + \cdot O_2^- + O_2 \quad\quad\quad (Eq.\ 4\text{-}6b)$$

$$\cdot O_2^- + H^+ \leftrightarrow HO_2 \cdot \quad\quad\quad (Eq.\ 4\text{-}6c)$$

$$O_3 + \cdot O_2^- \rightarrow \cdot O_3^- + O_2 \quad\quad\quad (Eq.\ 4\text{-}6d)$$

$$\cdot O_3^- + H^+ \rightarrow HO_3 \cdot \quad\quad\quad (Eq.\ 4\text{-}6e)$$

$$HO_3 \cdot \rightarrow \cdot OH + O_2 \quad\quad\quad (Eq.\ 4\text{-}6f)$$

The overall reaction is

$$H_2O_2 + 2\ O_3 \rightarrow 2\ \cdot OH + 3\ O_2 \quad\quad\quad (Eq.\ 4\text{-}6g)$$

The stoichiometric ratio of the reaction is 1 mol of H$_2$O$_2$ to 2 mol of O$_3$, and the mass ratio is 1 g of H$_2$O$_2$ to 2.82 g of O$_3$ or 1 g of O$_3$ to 0.354 g of H$_2$O$_2$. The efficiency of the initiation reaction (Eq. 4-6a) is increased at higher pHs. However, higher pHs can also affect by-product formation as discussed in chapter 6.

The Dark Fenton's Process

When ferrous salts and hydrogen peroxide are added to a water or wastewater near pH 3, the Fenton's reaction occurs (Eq. 4-4b) generating ·OH radicals. The Fenton's process

(Flaherty and Huang 1994, Wadley and Waite 2004) has been known since 1894 (Fenton 1894). This process is useful to reduce the TOC in highly contaminated waters (TOC >100 mg/L).

The Sonolysis Process

Exposure of a sample to high-frequency sound waves (20 kHz to 200 MHz) (ultrasound) is called *sonolysis* (Mason and Pétrier 2004). The trough of the sound pressure waves allows microscopic bubbles to grow in the solution. With the following peak of the pressure wave, the bubbles collapse adiabatically (no loss of heat), which creates very high temperatures (>2,000 K) in the collapsing bubbles. This causes some water vapor to dissociate into ·OH radicals and H atoms. Although degradation of pollutants with the sonolysis process has been demonstrated, the efficiencies are very low.

Radiation Processes

When high-energy particles (e.g., electrons) or radiation (e.g., γ-rays or X-rays) enter water, the high energy is transferred to the water resulting in many products including ·OH radicals, H atoms, and hydrated electrons (Cooper et al. 2004). In particular, the ·OH radicals can initiate the typical reactions of an AOT. The process is quite efficient; however, the capital costs can be quite high.

Wet Air Oxidation Process

In this process, polluted water is heated under a high pressure of oxygen to temperatures between 120°C and 300°C (Patria et al. 2004). Under these conditions, the dissolved oxygen at high temperature is able to oxidize and destroy the pollutants. This process can be quite expensive, both in operating and capital costs.

LIGHT-DRIVEN HETEROGENEOUS AOTS

Certain metal oxides (particularly the anatase form of TiO_2) absorb UV and generate ·OH radicals on the surface of the particles (Mills and Lee 2004). The relevant reactions are

$$TiO_2 + h\nu \rightarrow h^+_{TiO_2} + e^-_{TiO_2} \quad \text{(Eq. 4-7a)}$$

$$h^+_{TiO_2} + H_2O \rightarrow \cdot OH_{TiO_2} + H^+ \quad \text{(Eq. 4-7b)}$$

$$e^-_{TiO_2} + O_2 \rightarrow \cdot O^-_{2\,TiO_2} \quad \text{(Eq. 4-7c)}$$

The organic pollutant adsorbs to the surface of the TiO_2 particle and is then attacked by the adsorbed $\cdot OH_{TiO_2}$ radical. The quantum yield for reaction Eq. 4-7b is only ~0.04

(Sun and Bolton 1996); hence, this AOT suffers from a disadvantage in this "inefficient" step. However, there may be situations (e.g., small-scale municipal applications) where surface absorption of pollutants onto TiO_2 or the desire for a chemical free AOT may mitigate the relatively low quantum yield.

HOMOGENEOUS ADVANCED REDUCTION PROCESSES

Most advanced processes involve the generation of highly oxidizing radicals; however, there are a few that generate highly reducing intermediates.

UV/Iodide Process

The iodide ion absorbs UV in the 200–230 nm region and undergoes the photolysis reaction

$$I^- + h\nu \rightarrow I\bullet + e^-_{aq} \qquad \text{(Eq. 4-8)}$$

where e^-_{aq} stands for the hydrated electron, which is a very strong reducing agent. Bolton and Cater (1993) have shown that this process can degrade chlorinated alkanes, such as chloroform and carbon tetrachloride.

Photolysis of Sulfur-Containing Ions

Vellanki et al. (2013) have shown that the UV photolysis of several sulfur-containing ions can produce highly reducing species

$$S_2O_4^{2-} + h\nu \rightarrow 2\,SO_2^- \qquad \text{(Eq. 4-9a)}$$

$$SO_3^{2-} + h\nu \rightarrow SO_3^- + e^-_{aq} \qquad \text{(Eq. 4-9b)}$$

They report the reduction of target compounds, such as perchlorate, nitrate, perfluorooctanoic acid and 2,4-dichlorophenol.

SUMMARY

Table 4-1 compares the characteristics of the various AOTs that have been discussed in this chapter.

A wide variety of processes can be considered AOTs; however, the primary technologies installed at the municipal scale are UV/H_2O_2 and ozone/H_2O_2 AOTs. UV/chlorine has also gained interest in recent years with a few full-scale applications under design. Other technologies (e.g., UV/TiO_2) have fewer full-scale installations and, therefore, require detailed feasibility evaluations and testing prior to implementation. The remainder of the chapters focus primarily on UV/H_2O_2, UV/chlorine, and ozone/H_2O_2 as these are the most likely AOT selections for utilities.

Table 4-1 Comparison of AOTs

AOT	Relative Treatment Efficiency	Development Level	Driver or Examples
Direct photolysis	Depends on quantum yield of the target contaminant	Commercial	Treatment of NDMA
UV/O_3	Moderate	Research	Excited state of contaminant should react with O_3
VUV water photolysis	Depends on VUV lamp efficiency and absorbance of the water matrix	Research to commercial	Ultrapure water treatment in the semiconductor industry
UV/H_2O_2	High	Commercial	Treatment of micropollutants and taste-and-odor compounds
UV/chlorine	High	Research to commercial	Good for waters that absorb strongly below 300 nm and have a pH ≤6
UV-vis/Fenton	High	Research to commercial	Good for waters that absorb strongly below 300 nm but requires pH ~ 3
O_3/H_2O_2	High	Commercial	Good for waters that have a low UVT and low bromide concentrations
Dark Fenton's process	High	Commercial	Good for waters that have a high COD or TOC, but requires a pH ~ 3
Sonolysis	Low	Research	
Radiation processes	Moderate; high capital cost	Research	Good for waters with low UVT
Wet air oxidation	Moderate	Research	Good for waters with low UVT
UV/TiO_2	Low	Research to commercial	Better if contaminants adsorb to TiO_2. Sites that require a chemical free solution.
UV/iodide	Low	Research	Reduction process
Photolysis of sulfur-containing ions	Low	Research	Reduction process

REFERENCES

Beltrán, F.J. 2003. Ozone–UV Radiation–Hydrogen Peroxide Oxidation Technologies. In *Chemical Degradation Methods for Wastes and Pollutants—Environmental and Industrial Applications*. Tarr, M.A., Ed.New York: Marcel Dekker; e-edition, Taylor and Francis.

Bolton, J.R., and S.R. Cater, 1993. Treatment of Contaminated Waste Waters and Groundwaters With Photolytically Generated Hydrated Electrons, US Patent No. 5,258,124.

Bolton, J.R., and Cater, S.R. 1994. Homogeneous Photodegradation of Pollutants in Contaminated Water: An Introduction. In *Aquatic and Surface Photochemistry*. Helz, G.R., Zepp, R.G. and Crosby, D.G., Eds. Boca Raton, FL: Lewis.

Bolton, J.R., M.I. Stefan, P.-S. Shaw, and K.R. Lykke. 2011. Determination of the Quantum Yields of the Potassium Ferrioxalate and Potassium Iodide–Iodate Actinometers and a Method for the Calibration of Radiometer Detectors. *Jour. Photochem. Photobiol. A: Chem.*, 222:166–169.

Chan, P.Y., M. Gamel El-Din, and J.R. Bolton. 2012. A Solar-Driven UV/Chlorine Advanced Oxidation Process. *Wat. Res.*, 46(17):5672–5682.

Cooper, W.J., P. Gehgringer, A.K. Pikaev, C.N. Kurucz, and B.J. Mincher. 2004. Radiation Processes. In *Advanced Oxidation Processes for Water and Wastewater Treatment*, S. Parsons, Ed. London: IWA Publishing.

Feng, Y., D.W. Smith, and J.R. Bolton. 2007. Photolysis of Aqueous Free Chlorine Species (HOCl and OCl$^-$) With 254 nm Ultraviolet Light. *Jour. Environ. Eng. Sci.*, 6:277–284.

Fenton, H.J.H. 1894. Oxidation of Tartaric Acid in the Presence of Iron. *Jour. Chem. Soc. Trans.*, 65:899–910.

Flaherty, K.A., and C.P. Huang, 1994. Continuous Flow Applications of Fenton's Reagent for the Treatment of Refractory Wastewaters. In *Chemical Oxidation*. W.W. Eckenfelder, A.R. Bowers, and J.A. Roth, Eds. Lancaster, PA: Technomic Publishing.

Glaze, W.H., J.W. Kang, and D.H. Chapin. 1987. The Chemistry of Water Treatment Processes Involving Ozone, Hydrogen Peroxide and Ultraviolet Radiation. *Ozone Sci. Engin.*, 9(4):335–352.

Goldstein, S., and J. Rabani. 2008. The Ferrioxalate and Iodide–Iodate Actinometers in the UV region. *Jour. Photochem. Photobiol. A: Chem.*, 193:50–55.

Goldstein, S., D. Aschengrau, Y. Diamant, and J. Rabani. 2007. Photolysis of Aqueous H_2O_2: Quantum Yield and Applications for Polychromatic UV Actinometry in Photoreactors. *Environ. Sci. Technol.*, 41:7486–7490.

Heit, G, A. Neuner, P.-Y. Saugy, and A.M. Braun. 1998. Vacuum-UV (172 nm) Actinometry: The Quantum Yield of the Photolysis of Water. *Jour. Phys. Chem. A*, 102(28):5551–5561.

Mason, T.J., and C. Pétrier, 2004. Ultrasound Processes. In *Advanced Oxidation Processes for Water and Wastewater Treatment*. S. Parsons, Ed. London: IWA Publishing.

Mills, A., and S.K. Lee. 2004. Semiconductor Photocatalysis. In *Advanced Oxidation Processes for Water and Wastewater Treatment*. S. Parsons, Ed. London: IWA Publishing.

Nadtochenko, V.A., and J. Kiwi. 1998. Photolysis of FeOH^{2+} and FeCl^{2+} in Aqueous Solution. Photodissociation Kinetics and Quantum Yields. *Inorg. Chem.*, 37:5233–5238.

Oppenländer, T. 2003. *Photochemical Purification of Water and Air*. Weinheim, Germany: Wiley-VCH.

Patria, L., C. Maugans, C. Ellis, M. Belkhodja, D. Cretenot, F. Luck, and B. Copa. 2004. Wet Air Oxidation Processes. In *Advanced Oxidation Processes for Water and Wastewater Treatment*. S. Parsons, Ed. London: IWA Publishing.

Shu, Z., C. Li, M. Belosevic, J.R. Bolton, and M. Gamal El-Din. 2014. Application of a Solar UV/Chlorine Advanced Oxidation Process to Oil Sands Process-Affected Water Remediation. *Environ. Sci. Technol.*, 48(16):9692–9701.

Sun, L., and J.R. Bolton. 1996. Determination of the Quantum Yield for the Photochemical Generation of Hydroxyl Radicals in TiO_2 Suspensions. *Jour. Phys. Chem.*, 100:4127–4134.

Tarr, M.A. 2003. Fenton and Modified Fenton Methods for Pollutant Degradation. In *Chemical Degradation Methods for Wastes and Pollutants—Environmental and Industrial Applications*. M.A. Tarr, Ed. New York: Marcel Dekker; e-edition, Taylor and Francis.

Tuhkanen, T.A. 2004. UV/H_2O_2 Processes. In *Advanced Oxidation Processes for Water and Wastewater Treatment*. S. Parsons, Ed. London: IWA Publishing.

Vellanki, B.P., B. Batchelor, and A. Abdel-Wahab. 2013. Advanced Reduction Processes: A New Class of Treatment Processes. *Environ. Eng. Sci.*, 30(5):264–271.

Wadley, S., and T.D. Waite. 2004. Fenton Processes, in *Advanced Oxidation Processes for Water and Wastewater Treatment*, S. Parsons, Ed., London, UK: IWA Publishing.

Watts, M.J., and K.G. Linden. 2007. Chlorine Photolysis and Subsequent OH Radical Production During UV Treatment of Chlorinated Water. *Wat. Res.*, 41:2871–2878.

Weeks, J.L., G.M.A.C. Meaburn, and S. Gordon. 1963. Absorption Coefficients of Liquid Water and Aqueous Solutions in the Far Ultraviolet. *Rad. Res.*, 19(3):559–567.

5

Advanced Oxidation Equipment

Advanced oxidation equipment needs to be designed to provide adequate exposure of the target contaminant to the oxidizing molecules. The AOT system design parameters must account not only for the target contaminant oxidation but also for background scavenging ensuring that the contaminants and oxidizers have an effective contact time. Contact times are relatively short for UV-based AOTs but may be longer for ozone-based AOTs. The majority of AOT systems also incorporate chemical addition, and thus consideration must be paid to effectively inject the chemicals and, if necessary, subsequently quench any unused chemicals (e.g., hydrogen peroxide). AOT implementation requires consideration for the system (i.e., UV- or ozone-based AOTs) and the engineering requirements of the system including technology specifications, chemical injection and storage, and the necessary instrumentation and monitoring equipment required for the system. This chapter discusses the equipment associated with AOT systems and the key design considerations associated with the equipment.

AVAILABLE UV EQUIPMENT

If the AOT system uses UV light for the generation of hydroxyl radicals, the available UV equipment can be divided into two major types: *open-channel* and *closed-pipe* systems. The former are found primarily in wastewater treatment facilities and the latter in drinking water treatment facilities.

Open-Channel Systems

In open-channel systems, the water to be treated flows by gravity through a channel with a rectangular cross section with the water level maintained by a weir. The UV lamps are positioned either (or both) perpendicular or parallel to flow. In some open-channel systems, particularly those that use medium-pressure UV lamps, the water is forced to flow through circular or octagonal channels, which each contain a UV lamp and quartz sleeve. Open-channel AOT systems are not typically used in drinking water applications because UV treatment is usually applied postfiltration, where the water should not exposed to the environment. Open-channel systems may be applicable for wastewater application or for a limited number of drinking water applications.

Closed-Pipe Systems

In closed-pipe systems, the UV reactors consist of units with a circular or rectangular cross section that are inserted into the water flow (i.e., into a pipe) in a water treatment facility. The units contain UV lamps, UV sensors, quartz sleeves around the lamps, and may have a quartz-cleaning mechanism. There are many types of closed-pipe UV reactors but most can be classified into the following groups:

1. *Multiple lamp reactor with lamps parallel to flow* (Figure 5-1a). This is a cylindrical cross-section UV reactor with several lamps with quartz sleeves located parallel to the axis of the cylinder. This type of reactor is generally designed for a lower flow rate between 10 and 5,500 gpm. Most UV reactors of this type also use low-pressure UV lamps.

2. *Multiple lamp reactor with lamps perpendicular to flow* (Figure 5-1b). These lamps with their quartz sleeves traverse the axis of the reactor and are perpendicular to the water flow. The reactor cross section can be circular, square, or rectangular. The lamps can be oriented horizontally, vertically, or at some angle in between. This type of reactor may be able to handle higher flow rates depending on treatment goals (e.g., up to 14,000 gpm).

UV LAMPS

The most important components of the UV equipment are the UV lamps. These lamps must be able to generate UV light in the appropriate photochemical region (e.g., 200–300 nm) with acceptable efficiency and lifetime.

A UV lamp converts electrical energy partially into light energy and the remainder dissipates as heat. The lamp efficiency is usually defined as the ratio of the total UV power output in the germicidal region to that of the external electrical power input.[1]

Lamp Types

Several types of lamps that emit in the ultraviolet region are suitable for UV AOT applications (Table 5-1).

Currently, the most important lamps for UV AOT systems are gas discharge lamps. These usually have two electrodes, one at each end of a tube containing a gas, which may contain metal atoms (e.g., mercury) or a rare gas (e.g., xenon). In such lamps, light emission arises from atoms that have been excited to higher energy states by high-energy electrons emitted from the electrodes. There are several common types of gas discharge lamps.

The vast majority of lamps used in UV AOT systems are mercury vapor lamps. In these lamps, the electric current that flows through the ionized hot plasma causes the

---

[1] Some manufacturers use the *electrical power to the lamp* as the basis for efficiency specifications. This definition neglects the power losses in the power supply.

Advanced Oxidation Equipment 63

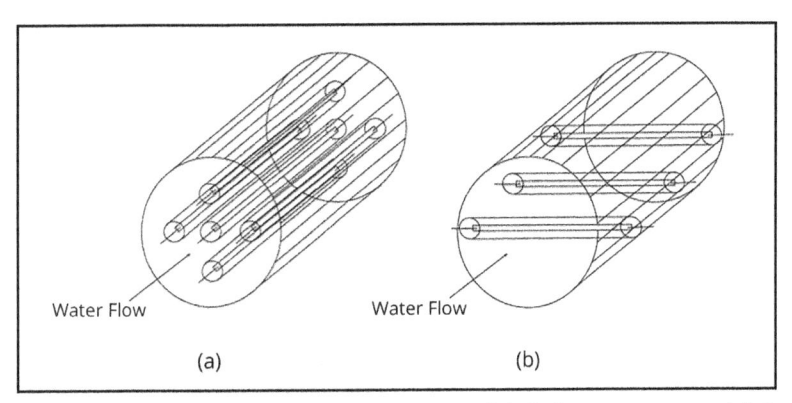

Figure 5-1 Closed-pipe UV reactor types: (a) multiple lamp reactor with lamps parallel to flow and (b) multiple lamp reactor with lamps perpendicular to flow

Table 5-1 Classification of UV lamps

Lamp Type	Characteristics	Wavelength Range	Example
Gas Discharge Lamps*	Electrodes at each end of a long cylindrical lamp containing a metal vapor (e.g., mercury)	150–800 nm	Fluorescent light bulb; low-pressure and medium-pressure mercury lamps
Light-Emitting Diodes (LEDs)	Solid-state semiconductors that emit light when current is passed through	250–600 nm	Indicator lights on electronic equipment
Excilamps	Electric discharge in a lamp containing an inert gas (e.g., Xe) or a halogen (e.g., Cl_2)	Relatively monochromatic in the range 170–300 nm	New experimental type of lamp

*Currently the most commonly used lamps for AOTs.

mercury atoms to be raised to higher energy excited states. When these excited states return to their ground state, UV light is emitted with a wavelength inversely proportional to the energy difference between the excited state and the ground state.

The lifetime of the excited mercury atoms is limited to microseconds or milliseconds; they decay to lower energy levels with the excess energy emitted as light or heat. The wavelength of the light is determined by the difference in the energy levels (see Eq. 2-1b). The relative intensity of the various transitions varies considerably.

There are three common types of mercury discharge lamps used in AOT applications: low-pressure high-output (LPHO), LPHO-amalgam, and medium-pressure (MP) lamps.

Low-Pressure High-Output Mercury Amalgam Lamps

Since about 1998, two types of modified low-pressure lamps have been introduced. The first type is called an LPHO lamp. It consists of a standard low-pressure (LP) lamp but with reinforced filaments standing out 40–60 mm from the end of the lamp. This configuration allows the lamp to maintain a current 1.5–2 times that of a standard LP lamp without heating the electrodes too much. If the electrodes get too hot, the mercury vapor pressure rises and the overall efficiency drops.

A more recent type of enhanced low-pressure lamp is the amalgam lamp.[2] These lamps have heavy electrodes and contain no free mercury. A solid amalgam (a compound of mercury with another element, such as indium or gallium) spot is placed on the inner wall of the lamp. This spot serves to control the mercury vapor pressure, even though a higher current is flowing. Amalgam lamps can have a UVC output 2 to 3 times that of a conventional LP lamp.

Both LPHO and amalgam lamps are similar to conventional LP lamps in that they have a primary emission at 253.7 nm but fewer lamps are required due to the ability to handle higher currents. The higher operating currents result in fewer lamps required for a given level of treatment. LPHO and amalgam lamps also operate at temperatures around 100°C, which allows them to be less sensitive to the water temperature than LP lamps.

Medium-Pressure (MP) Mercury Lamps

MP mercury lamps emit light between 200 and 400 nm. The emission lines of a mercury lamp are only sharp when the pressure of the gas is low (<100 kPa). If the pressure is increased, the lamp can carry much more power, but the increased collision rate from other gas molecules causes the emission lines to broaden. For the same length of lamp (about 120 cm), an MP lamp (pressure about 100 kPa or about 1 atm) can carry up to 30,000. Figure 5-2 shows a comparison of the emission of low-pressure and MP lamps in the ultraviolet region. MP lamps operate very hot; thus they are not too sensitive to the ambient temperature.

Flash Lamps

Flash lamps are similar to continuous wave (CW) arc lamps in that they consist of a cylindrical quartz tube with electrodes at each end and filled with a gas (e.g., xenon). A power supply *fires* the lamps by discharging a large amount of electrical energy (100–1000 J) in a very short period of time (several μ sec) by applying a very high voltage (10–30 kV). The resulting plasma reaches temperatures of 8,000–25,000 K, and the emission is essentially that of a black body.[3] In some experimental flash lamp systems, the lamp is flashed

---

2 Some manufacturers often also call these lamps low-pressure high-output lamps.
3 All objects at temperatures above 0 K emit light, called *black-body* emission. The peak wavelength shifts to shorter wavelengths as the temperature increases.

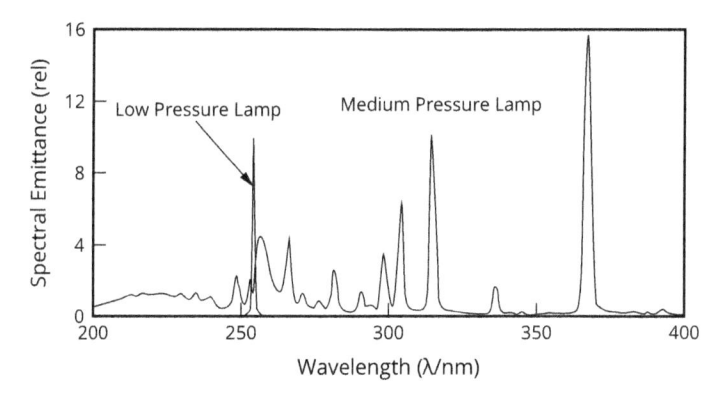

Figure 5-2 Relative spectral emittance from low-pressure and medium-pressure lamps

Table 5-2 Emission wavelengths for some common excilamps (Oppenländer 1994, 2012)

Excimer	Wavelength (λ/nm)	Excimer	Wavelength (λ/nm)
Xe_2^*	172	$XeBr^*$	282
$KrCl^*$	222	$XeCl^*$	308
XeI^*	253	I_2^*	342
Cl_2^*	259		

* Indicates that the entity is in an electronically excited state.

about 30 times per second. At present, the germicidal efficiency and lifetimes are low, and very few commercial systems use flash lamps.

Excilamps

Excilamps (also called *excimer lamps*) are unique in that they emit in a narrow band of wavelengths. They involve the formation of excimers, which are molecular dimers that are stable only in the excited state and dissociate when decaying to the ground state. Table 5-2 gives the wavelengths of some of the common excimer lamps. Oppenländer (1994) and Sosnin et al. (2006) describe these lamps and their applications. Excilamps are currently only in the research stage, but if the efficiencies and lifetimes improve significantly, they could become viable commercially.

Comparison of Lamp Types

Table 5-3 gives a comparison of the characteristics of the principal lamp types that are used in UV disinfection reactors. It should be noted that LPHO and amalgam lamps have the highest germicidal efficiency. However, the overall efficiency for an AOT application depends on the molecular adsorption spectra of the target oxidant (e.g., hydrogen

Table 5-3 Comparison of the characteristics of UV lamps used for UV disinfection of drinking water

Characteristic	LPHO*	Medium Pressure
Emission	Virtually monochromatic (253.7 nm)	Polychromatic (185–600 nm)
Mercury vapor pressure (Pa)	0.1 – 10 Pa	50–300 kPa
Mass of mercury for a 1.2-m long lamp	~30–75 mg	2–4 g
Operating bulb temperature	60–100°C	600–900°C
Arc length (cm)	15–200	10–200
Lifetime	7,000–12,000 hours	3,000–9,000 hours
Power density (input power in W/cm)	0.6–1.2	125–200

* LPHO lamps are either amalgam lamps or lamps with special ends that allow a cold spot at the ends of the lamp.

peroxide or chlorine). Depending on water quality, MP lamps could be more efficient as hydrogen peroxide has a higher adsorption at wavelengths below 254 nm. The optimal lamp selection is a site-specific selection based on water quality, target contaminant, power costs, and annual operations and maintenance considerations.

UV SENSORS

The UV sensor is an important component of a UV reactor because each UV sensor monitors the irradiance (or intensity) at specific locations in the UV reactor. Depending on the control strategy for the UV reactor, the UV sensor readings may be used to calculate the UV dose (fluence) delivered by the reactor and thus become an important part of the online monitoring system. Some AOT systems that operate based on a target E_{EO} may not directly use the UV sensor readings as treatment is calculated based on power input.

Most UV sensors use a light-sensitive semiconductor (e.g., SiC) that is *solar blind*. This means that the sensor is not activated by solar light, which has wavelengths >300 nm. Often a filter is applied so that the sensor is active primarily in the germicidal range of about 200 – 300 nm. Most current UV sensors do not monitor wavelengths below 240 nm and these wavelengths can be important in MP UV reactors for hydroxyl radical formation. UV sensors are in development to monitor the lower wavelengths in disinfection applications and should be considered for MP UV AOT applications, if the low-wavelength sensors are proven to be viable options.

UV sensors continuously monitor the irradiance at a specific position in the UV reactor; however, the sensor reading is affected by several factors: (1) the UV output of the lamp, (2) the transmittance of the quartz sleeve (fouling of this sleeve can drastically decrease the UV light that enters the reactor), (3) the transmittance of the water (an increase in absorbing substances in the water will decrease the sensor reading), and (4) the transmittance of the sensor window (again, fouling of this surface can drastically reduce the sensor reading). Proper maintenance of the UV reactor should minimize the effects of factors (2) and (4), so that factors (1) and (3) are the most significant.

If the sensors are used to control UV reactor operations, it is recommended that the UV sensors be checked against a reference sensor periodically, and the reference sensor should be calibrated periodically (once per year). Reference sensor checks are a regulatory requirement for disinfection credit but are not required for AOT-only applications. Sensor calibration checks are performed using a reference sensor that is identical to the duty sensor, except that it has been calibrated against National Standards, such as those maintained by the National Institute for Standards and Technology (NIST). This verifies that the UV sensors are reading irradiances accurately.

SLEEVES

A UV lamp must be separated by an air space from the flowing water because a UV lamp needs to operate at a higher temperature than the water. This separation is accomplished by placing the lamp inside a sleeve, which is usually made of natural or synthetic quartz because quartz is one of the few materials that transmits UV in the 200–300 nm region.

Cleaning of Sleeves

Sleeves can get fouling and deposits on the surface. Most mineral salts (e.g., calcium, magnesium, and iron salts) have a solubility that decreases with increasing temperature. Therefore, the quartz sleeve, which is close to a hot UV lamp, has a higher temperature than the surrounding water providing a place where these salts may precipitate. This fouls the quartz sleeve and decreases the UV transmittance. Therefore, it may be necessary to design UV reactors with an effective quartz cleaning mechanism depending on the site-specific water quality. In small UV reactors, it may be sufficient to introduce a maintenance schedule that includes periodic removal. In larger UV reactors, an active quartz cleaning system may be necessary. This can either be mechanical (e.g., using a stainless-steel brush) or chemical (e.g., using a sealed acid solution) or a combination of both methods. Periodically, the cleaning system should move across the quartz sleeve to clean off any deposits that may have formed. Some larger UV reactors can also have off-line chemical cleaning systems that spray or fill an acid solution into the reactor. These systems require the reactor to be taken off-line, but they do not require the lamp sleeves to be removed for cleaning.

OTHER COMPONENTS

Several other components are essential to the operation of a UV reactor.

Lamp Power Supplies

The lamp power system is usually external from the UV reactor itself and provides the proper voltage and current to operate each UV lamp. Most UV lamps require a ballast to provide the starting voltage and to stabilize the current during operation. This is usually done passively using an inductor; however, there now exist electronic ballasts that actively control the operation of the UV lamps and may operate at a much higher frequency than the line frequency. The power supply exhibits a certain efficiency (about 92–95 percent for inductive ballast power supplies and 95–98 percent for electronic power supplies), defined as the ratio of the power across the lamp to the power drawn from the wall.

Control of UV Reactors

Several factors need to be monitored such that operators can be assured that the UV reactor and chemical feed systems are operating in a safe range to achieve the target level of treatment. The most important readings are UV sensor readings or power input, flow rate, and UV transmittance (UVT at 254 nm for a 1 cm path length) using an online transmittance monitor. Some manufacturers have developed an algorithm that can be programmed into a programmable logic controller (PLC) that takes the inputs of flow rate and UV transmittance and provides the necessary oxidant dose and UV lamp power setting needed to achieve the target level of treatment. The PLC will also monitor the UV reactor and chemical feed system operation and alarm statuses.

AVAILABLE OZONE EQUIPMENT

If the AOT system uses ozone for the generation of hydroxyl radicals, the system will require several components. The following section provides a brief introduction to the various components of an ozone system.

Ozone Generators

Ozone decays rapidly during storage and cannot be stored; therefore, ozone must be generated where the AOT is occurring. Ozone generators produce ozone from either liquid oxygen (LOX) or onsite oxygen separation techniques including pressure swing adsorption (PSA) or vacuum pressure swing adsorption (VPSA). LOX systems have grown to be the preferred method of ozone generation for many utilities as they are less complex and can be a more economically favorable method of ozone generation depending on LOX availability. However, some locations have selected PSA systems to reduce chemical storage and handling requirements. Earlier generations of ozone systems utilized air preparatory systems and were air fed but produced ozone at a lower concentration than current technology.

The basic concept of ozone generation involves breaking oxygen molecules (O_2) into oxygen atoms (O), which then combine with other oxygen molecules to form ozone (O_3). Within ozone generators, oxygen molecules are split by sending oxygen containing gas through a small gap with electrical energy flowing across the gap. The electrons flowing across the electrical gap split the oxygen molecules. The concentration of ozone produced from ozone generators varies based on the operating conditions of the generator and can often be optimized to site-specific needs.

There are multiple types of ozone generators that can be considered. Two examples are the conventional ozone systems utilizing a tube and shell generator that are supplied by most major manufacturers and modular ozone generation systems that are an emerging technology. Modular ozone generators are made up of multiple standard-sized cells combined into cabinets to meet the projects production needs. The modular systems have the operational flexibility of adding more cells to the generator to meet higher ozone demand rather than requiring additional generators as seen with conventional ozone generation.

Ozone Contactors

The ozone and hydrogen peroxide in an AOT system require sufficient contact time to generate the hydroxyl radicals and achieve the target level of treatment. The amount of contact time will depend on multiple factors including

- Disinfection CT requirements (if applicable)
- Ozone dose
- Ozone demand and decay characteristics
- Target contaminant and treatment level
- Location of peroxide addition (preozone or postozone addition)

Contactor designs can vary depending on the facility size and site constraints. Two main contactor design options include pipeline contactors and large engineered contact chambers. Pipeline contactors are typically limited to smaller flows. Pipeline contactors can provide a smaller footprint but may have higher head loss due to additional fittings and potentially in-line static mixers.

During the selection and design of ozone contactors, the following items should be considered:

- Provide adequate hydraulic residence time
- Minimize short circuiting and maximize baffling factor
- Minimize quenching requirements (ozone or peroxide) at last cell of contactor
- Minimize head loss through facility
- Size cells to allow sampling of residual ozone (if required for disinfection credit)

Ozone Destruct Units

Most ozone designs will have an ozone transfer efficiency of 90–95 percent. This indicates that 90–95 percent of the added ozone is dissolved into the water. The remaining ozone has the potential to off-gas, which can be a health and safety risk. As a result, ozone systems include ozone destruct unit that capture the ozone gas (generally through maintaining a vacuum on the contactor) and pass it through a destruction catalyst before discharging the air to the atmosphere.

Liquid Oxygen Storage and Vaporizers

LOX is stored in cryogenic storage tanks under pressure before it is used as feed gas in the ozone generation process. A small fraction of the LOX in the storage vessel will boil off and is managed by an economizer loop that allows the vapor (gaseous oxygen) to feed directly out of the LOX tank to the process into the header feeding the ozone generators.

Ambient vaporizers use surface area of the vaporizer to promote the heat exchange needed to vaporize the LOX to form gaseous oxygen. Vaporizers should be sized to handle the full range of operation of the installed ozone generator capacity.

Nitrogen Boost

A nitrogen boost skid with air compressors can be added to the system to provide nitrogen gas (typically 1–2 percent) to the gaseous oxygen stream to stabilize the conversion reaction of oxygen to ozone. This results in a higher ozone concentration and prolongs the life of the ozone generation components (dielectrics) in the ozone generators. The nitrogen source is typical ambient air due to the nitrogen concentration in air.

Ozone Monitoring

Ozone monitoring is critical for system control and operator safety. Ozone residual analyzers are used to monitor the ozone residual in the water at locations throughout the contactor. Residual analyzers are recommended to identify the residual established upon ozone addition and track the residual for CT calculation for disinfection credit. The number of ozone residual analyzers will depend on the contactor design and need for disinfection credit. The more analyzers that are installed, the more accurately the decay through the contactor can be monitored. However, ozone residual analyzers are maintenance intensive and number installed should consider the time required for maintenance and calibration.

Ambient ozone analyzers should be used to monitor for ozone leaks that could endanger operator safety. Ambient oxygen analyzers should be provided throughout the facilities where oxygen and ozone piping is located.

Control of Ozone System

There are multiple control approaches available for operating an ozone system. One approach is to manually set the ozone dose and allow the system to produce sufficient ozone to achieve that dose based on the current plant flow. In this scenario, the dose setpoint could be based off of operational experience or bench/pilot data to effectively accomplish the intent of the system.

Advanced Oxidation Equipment

Table 5-4 Ozone instrumentation

Analyzer	Purpose
Dewpoint	Ensure that the oxygen feed to ozone generators is dry
Ambient Oxygen	Safety monitor to detect oxygen leaks
Ambient Ozone	Safety monitor to detect ozone leaks
High Concentration Ozone	Monitor on generator to identify concentration of ozone being produced by generator
Off-gas Ozone	Monitor to measure concentration of ozone in off-gas from contactor and being discharged from destruct units
Residual Ozone	Analyzer to determine concentration of ozone in water; can be used to verify proper process operation and evaluate demand and decay characteristics

Alternatively, the ozone system could be set to operate based on feedback from residual ozone analyzers to achieve a setpoint ozone residual in the water. The system will then vary the ozone dose to achieve a target residual in the water.

There are many instruments involved within an ozone system to ensure proper equipment operation and safety. Table 5-4 presents an overview of ozone instrumentation. In addition to the instruments in Table 5-4, there are also pressure and temperature instruments on the system to verify proper equipment operation.

ANCILLARY AOT EQUIPMENT
Oxidant Chemical Systems

Most UV and ozone AOTs require the addition of an oxidant (e.g., hydrogen peroxide or chlorine) for radical formation. The oxidant is dosed into the system prior to the UV system and prior or during ozone exposure. Hydrogen peroxide can be added prior to ozone to limit bromate formation; however, this type of system cannot achieve disinfection credit. If disinfection credit is desired, the hydrogen peroxide is added after the required CT for disinfection is achieved. The oxidant dosing system generally consists of a storage tank, metering pumps, and associated piping.

Oxidant dosing requirements are based on the operating conditions and water quality. UV/peroxide AOTs generally require more hydrogen peroxide than ozone-based AOT, as hydrogen peroxide has a relatively lower molar adsorption coefficient.

Hydrogen Peroxide Quenching System

Excess hydrogen peroxide must be quenched if a disinfectant residual is to be maintained downstream of the AOT system. Any residual hydrogen peroxide can be quenched chemically or catalytically through a reactor with granular activated carbon (GAC). The majority of AOT systems that require quenching use chemical quenching with sodium hypochlorite as it has a relatively low stoichiometric dose and is a chemical commonly used in most water systems. However, consideration needs to be given to chemical reaction rates, adequate mixing, dose pacing, and by-product concerns when using sodium hypochlorite. In recent years, a number of systems have successfully implemented GAC quenching.

REFERENCES

Oppenländer, T. 1994. Novel Incoherent Excimer UV Irradiation Units for the Application in Photochemistry, Photobiology/Medicine and for Waste Water Treatment. *EPA Newsletter*, March 1994, pp. 2–8. Zürich, Switzerland: European Photochemistry Association.

Oppenländer, T. 2012. Excilamp Photochemistry. In *CRC Handbook of Organic Photochemistry and Photobiology*, 3rd Ed. A. Griesbeck, M. Oelgemöller, and F. Ghetti, Eds. Boca Raton, FL: CRC Press.

Sosnin, E.A., T. Oppenländer, and V.F. Tarasenko. 2006. Applications of Capacitive and Barrier Discharge Excilamps in Photoscience. *Jour. Photochem. Photbiol. C: Photochem Reviews*, 7:145–163.

6

Effects of Water Quality on AOT Systems[1]

Water quality can have a significant effect on the design and performance of AOT systems. Water quality effects can vary based on the specific AOT. For example, UV transmittance (UVT) is a major driver for UV AOT system sizing and efficiency; however, UVT will not have a direct effect for an ozone AOT system. The water quality characteristics that affect AOTs include the same water quality parameters that affect UV or ozone disinfection applications, but additional consideration must be given to water quality parameters that affect hydroxyl radical scavenging and by-product formation. This chapter will discuss the importance of water quality and its effects on design and performance.

UV TRANSMITTANCE AND ABSORBANCE

As with disinfection applications, UVT is an important parameter for UV AOT systems. UVT is a measure of how much UV light is absorbed by the background constituents in the water, which would then not be available to react with the target contaminant (photolysis applications) or oxidants (e.g., hydrogen peroxide or chlorine). As the UVT decreases, more UV light is absorbed and thus more energy (i.e., more lamps) is needed to deliver the necessary UV energy or UV dose for achieving the target level of contaminant reduction. Typically, UVT at 254 nm is used as the main water quality design parameter; however, the UVT spectrum (i.e., UVT scan) is equally important for UV reactors that use polychromatic light sources (e.g., MP lamps) because wavelengths other than 254 nm affect the system performance. For example, hydrogen peroxide adsorbs more strongly at wavelengths below 254 nm and understanding the UV absorbance at the lower wavelengths is necessary for proper sizing and operation.

---

[1] Some of this chapter has been adapted (with permission) from Bolton and Cotton (2008).

UVT at 254 nm

UVT_{254} is expressed as a percentage and defined as

$$\text{UVT} = 100 \times \left(\frac{E^{\ell}}{E^{0}}\right) = 100 \times 10^{-A_{254}} \quad \text{(Eq. 6-1a)}$$

where E^0 and E^{ℓ} are the irradiances incident on the cell and transmitted through a path length ℓ (the internal thickness of the cell), respectively, and A_{254} is the absorbance (actually absorption coefficient) at 254 nm for a 1 cm path length. In cases where the UVT is higher than 85 percent (often the case with many drinking waters), a 5- or 10-cm cell should be used to measure the absorbance ($A = a\ell$) for increased accuracy. The absorption coefficient (or A_{254}) is obtained by dividing the measured absorbance by the path length ℓ. Because the UVT is defined only for a path length of 1.00 cm, the UVT should be determined from the absorption coefficient a, that is

$$\text{UVT} = 100 \times 10^{-a} \quad \text{(Eq. 6-1b)}$$

A variety of absorbing components can result in increased absorption coefficients resulting in decreasing UVT. Total organic carbon (TOC), which consists of humic and/or alginic acids and other possible components, can be a major contributor to an increased absorption coefficient (i.e., decreasing UVT). Generally, the absorption coefficient of drinking waters increases as the TOC increases. Other components can also contribute to decreases in UVT. Some of these components are listed in Table 6-1.

When collecting and analyzing UVT samples, the water sample should not be filtered or the pH adjusted, which is suggested in *Standard Methods* (APHA, AWWA, and WEF 2012), because this could bias the UVT results.

UV Absorbance Spectrum (200–300 nm)

The dependence of the required UV energy or UV dose on the UV absorbance spectrum of the source water is different for UV AOT systems using medium-pressure UV lamps, as opposed to systems using low-pressure high-output UV lamps. The UV emission from medium-pressure UV lamps has a broad distribution over the wavelength range of 200–300 nm (see chapter 5). The UV emission from low-pressure high-output lamps is essentially monochromatic at 254 nm. The UV absorbance spectrum of natural waters generally increases with increases in the TOC level. However, the overall UV absorbance spectrum, particularly at the lower wavelengths, can be heavily influenced by other water constituents, such as nitrate.

For UV AOT systems using MP UV lamps, the situation is more complex than for monochromatic systems. To evaluate the implications of the full UV absorbance spectrum, information should be obtained about the emission spectrum of the UV lamps, the UV absorbance spectrum of the water, and the wavelength dependent adsorption of

Table 6-1 Summary of the molar absorption coefficients at 254 nm for components that may be present in drinking water

Compound	Molar Absorption Coefficient (M^{-1} cm^{-1})	Absorption Coefficient (cm^{-1}) for a 1 mg/L solution	pH
Ammonia (NH_3)*	~0	0	11.5
Ammonium (NH_4^+)*	~0	0	7.0
Calcium ion (Ca^{2+})*	~0	0	6.5
Ferric [$Fe(OH)^{2+}$]*	4,716	0.084	4.0
Ferrous (Fe^{2+})*	28	0.00050	5.0
Hydrogen peroxide (H_2O_2)*	18.7	0.00055	7.0
Hydroxide ion (OH^-)*	~0	0	13.3
Hypochlorite (ClO^-)†	65.7	0.0019	10.0
Hypochlorous acid ($HOCl$)†	59.0	0.0017	5.0
Magnesium ion (Mg^{2+})*	~0	0	6.0
Manganous ion (Mn^{2+})*	~0	0	3.6
Ozone (O_3) (aqueous)*	3,250	0.068	~7
Permanganate ion (MnO_4^-)*	657	0.012	5.5
Phosphate ion species ($H_2PO_4^-$, HPO_4^-)*	~0	0	5–9
Sulfate (SO_4^{2-})*	~0	0	7.0
Sulfite (SO_3^{2-})*	16.5	0.00052	9.0
Thiosulfate ($S_2O_3^{2-}$)*	201	0.0063	5.0
Zinc ion (Zn^{2+})*	1.7	0.000026	5.0

* Bolton et al. 2001.
† Feng et al. 2007.

the target contaminant or oxidant. Polychromatic lamps have the potential to increase the efficiency of AOT systems (e.g., generation of hydroxyl radicals) because of the increased absorbance of hydrogen peroxide at lower wavelengths (Figure 3-11), but the benefits are dependent on the UV absorption spectrum of the water, emission spectrum of the lamps, and efficiency of the lamps. Individual calculations should be performed for each specific case.

Estimating the difference in performance of a UV/H_2O_2 AOT driven by LPHO UV lamps (monochromatic at 254 nm) versus the same system driven by MP UV lamps is important. This can be done by calculating the fraction of light absorbed by H_2O_2. This is easily for the LPHO-driven system by just applying the Beer-Lambert Law at 254 nm. However, the process is more complex for the MP-driven system. Appendix C illustrates how to perform this calculation.

HYDROXYL RADICAL SCAVENGING DEMAND

Oxidation by hydroxyl radicals is nonspecific, which results in a background scavenging demand for the radicals in the source water. The background scavenging demand must be accounted for to achieve the target level of treatment. Water quality parameters that influence the background scavenging demand include, but are not limited to, carbonate species (i.e., alkalinity), natural organic matter (NOM), and other contaminants. Appendix B presents the hydroxyl radical rate constants for some of the principal water components that contribute to the hydroxyl radical scavenging demand. The hydroxyl radical rate constants for the common scavengers (i.e., carbonates and NOM) are slower than for most target contaminants, meaning the reactions occur slower with the scavengers. However, the higher concentrations of the scavengers (concentrations typically in mg/L versus ng/L to μg/L for the target compounds) result in the scavengers being an important water quality consideration and can impact equipment sizing and capital and operational costs.

High alkalinity waters can have a large hydroxyl radical scavenging demand. Carbonate and bicarbonate are two principal ions involved in hydroxyl radical scavenging with carbonate having a higher reaction rate constant than bicarbonate (appendix B). Typical drinking water pH values (~6–8) favor the bicarbonate species. A higher pH will result in increased concentrations of carbonate ions.

Similarly, free chlorine concentrations can add to the hydroxyl radical scavenging demand and will be pH dependent. Hypochlorous acid has a pK_a of 7.6. At a pH above 7.6, the hypochlorite ion is the predominant chlorine species. The hydroxyl radical rate constant for hypochlorite ion is four orders of magnitude higher than hypochlorous acid (appendix B) and comparable to the hydroxyl radical rate constants of many target contaminants. Due to the concentration of active chlorine, the speciation of chlorine can have a large impact on the efficiency on an UV/chlorine AOT system (Kwon et al. 2014a).

NOM can also effectively scavenge hydroxyl radicals (Keen et al. 2014); however, its effect is much less well understood and less predictable than carbonate and bicarbonate. The unpredictability of NOM is because of the variability in NOM composition. NOM has a highly variable structure, is often inconsistent from comparable sources, and can fluctuate seasonally. Therefore, designers of AOT systems must thoroughly examine the background scavenging properties of their source water to ensure optimal system performance.

The background scavenging demand can be determined using models with assumed hydroxyl radical constants for NOM and other water quality parameters. However, the assumed hydroxyl radical rate constants may not be accurate for all water qualities, as the nature of NOM varies among source waters. Alternatively, bench-scale methods have also been developed to directly calculate the scavenging demand of a water (Rosenfeldt and Linden 2007, Kwon et al. 2014b). Currently, the scavenging demand measurements are not standardized and are mainly only available through research laboratories or from manufacturers.

Research is also being conducted on a potential portable methods that could be used to evaluate the hydroxyl radical scavenging demand in the field (Rosenfeldt 2011, Donhame et al. 2014). Evaluating the hydroxyl radical scavenging demand is critical for properly sizing an AOT system and can also be useful for identifying the impact of water quality changes on scavenging demand. Waters with a low scavenging demand can be treated more cost-effectively using AOT technologies because chemical addition and power costs required to overcome the background scavenging demand are reduced.

pH

Source water pH can have an effect on the AOT process in several different ways. As discussed in the previous section, the pH can have a large influence on the hydroxyl radical scavenging demand by determining the speciation of key scavengers (e.g., carbonates or chlorine). The source water pH can also influence ozone decay rates and by-product formation in ozone AOTs. Ozone decay rates are influenced by the pH of the water with ozone decay rates increasing at higher pH values. As a result, changes in pH and subsequent changes in ozone decay rate can alter the dynamics and efficiency of hydrogen peroxide reactions and hydroxyl radical formation.

As with carbonates and chlorine, the speciation of hypobromous acid is an important parameter in determining bromate formation and is discussed in further detail in the following by-products section.

The quantum yield of a compound can also be pH dependent, which is important for UV photolysis applications. NDMA is an example compound that is often treated through UV photolysis. The quantum yield for NDMA increases as pH decreases (Lee et al. 2005). This results in UV photolysis being more energy efficient at lower pHs.

TURBIDITY

Turbidity is caused by suspended particles in the water. These particles can affect the UV light distribution in the water, either by scattering the UV light or by absorption of UV light by components in the particles. In the case of scattering, the UV light is not lost; it is just redirected, and the effect on UV performance is minimal. However, in the case of particle absorption, the UV light is lost, and this can adversely affect UV reactor performance. Disinfection applications can be negatively impacted by high turbidity (i.e., >5 ntu) as turbidity can shield microorganisms from the UV light. Turbidity generally has a limited influence on the AOT applications at turbidity levels typically found in potable water applications (i.e., <5 ntu).

DISINFECTANT RESIDUAL

The presence of a free chlorine residual can have an adverse effect on AOTs using hydrogen peroxide. Free chlorine will react and quench part of the hydrogen peroxide residual.

Excess hydrogen peroxide addition may be necessary to achieve the target influent concentration if the water contains a free chlorine residual. The reaction rate between free chlorine and hydrogen peroxide is 2.09 mg/L Cl_2/mg/L H_2O_2 (Liu et al. 2003). Chloramines have a much slower reaction rate with hydrogen peroxide as compared to free chlorine. As a result, a chloramines residual would not need to be quenched prior to a peroxide-based AOT. However, breakpoint chlorination would be required to achieve a free chlorine residual for a UV/chlorine AOT, if a chloramines residual is present.

For ozone AOTs, chlorine and chloramines residual can poison the ozone destruction catalyst, which is responsible for treating the ozone off-gas. Catalyst poisoning will shorten the life of the catalyst and increase the replacement frequency. The extent of ozone catalyst poisoning depends on the concentration of chlorine or chloramines; however, concentrations typical of disinfection residuals can significantly shorten the life of the catalyst. The presence of chlorine or chloramines should be considered when designing and maintaining an ozone destruct system.

UV/chlorine AOTs would not require quenching of a free chlorine residual as free chlorine is the target oxidant. However, the free chlorine concentration should be closely monitored. High free chlorine concentrations can result in pitting of the stainless steel within the reactors. The treated-water chlorine concentration must also be monitored as free chlorine has a USEPA maximum residual disinfectant level (MRDL) of 4 mg/L.

UV LAMP SLEEVE FOULING

UV lamp sleeve fouling can be affected by a range of water quality parameters and exact predictions of fouling rates are difficult without site-specific fouling studies. Depending on the water quality, fouling can occur in hours or months. Water quality parameters that have been shown to influence fouling include hardness (as $CaCO_3$), alkalinity, temperature, ion concentration, oxidation reduction potential (ORP), and pH. Fouling can occur for the following reasons:

- Compounds for which the solubility decreases as temperature increases may precipitate (e.g., the carbonate, sulfate, and phosphate salts of Mg^{2+}, Ca^{2+}, Fe^{3+}, and Al^{3+}).
- Photochemical reactions that are independent of sleeve temperature may cause sleeve fouling (Derrick and Blatchley 2005).
- Compounds with low solubility may precipitate [e.g., $Fe(OH)_3$ and $Al(OH)_3$].
- Particles may deposit on the lamp sleeve surface arising from gravity settling and turbulence-induced collisions (Lin et al. 1999).
- Organic fouling can occur when a reactor is left off and full of water for an extended period of time (Toivanen 2000).
- Inorganic constituents can oxidize and precipitate (Wait et al. 2005).

Upstream chlorination has been shown through pilot and full-scale experiences to increase the fouling potential of a water. AOT systems using chlorine as the oxidant should consider the impact of chlorine on the fouling potential of the water and should consider potential online or off-line cleaning systems.

BY-PRODUCTS FROM AOT TREATMENT

As with any oxidation process, unintended by-product formation is possible as a result of AOT treatment. The possible by-products vary by technology and are dependent on source water quality and other operating conditions of the system (i.e., chlorination). By-products include both regulated and unregulated compounds. The following sections summarize some of the known potential by-products associated with AOTs.

Regulated By-products

AOTs have the potential to form or increase the formation potential of several regulated by-products including disinfection by-products (DBPs), nitrite, and bromate.

Disinfection By-products

DBPs are a concern for systems using chlorine as a residual disinfectant. Regulated DBPs include trihalomethanes (THMs) and haloacetic acids (HAAs). In UV disinfection systems, UV doses are relatively low and there are no significant changes in regulated DBP levels following doses typical of UV disinfection treatment (Malley et al. 1996, Kashinkunti et al. 2003, Zheng et al. 1999a, Liu et al. 2002, Venkatesan et al. 2003). However, research has shown mixed results at the higher doses typical in AOP applications. Dotson et al. (2010) and Venkatesan et al. (2003) showed that UV doses of greater than 400 mJ/cm^2 increased the THM and HAA formation potentials. TTHMs and HAA5 are regulated under the Stage 1 and Stage 2 Disinfectants/Disinfection By-product (D/DBP) Rule. However, Lui et al. (2002) and Zheng et al. (1999b) showed little to no increases in TTHMs and HAAs at similar doses indicating the impacts of UV photolysis on TTHM and HAAs may depend on the water quality.

AOTs have also been shown to transform organics in the water, which can result in higher formation of DBPs, including TTHMs and HAA5, after free chlorine contact time (Zheng et al. 1999b, Dotson et al. 2010, Andrews 2009, Wang et al. 2015). The increase in DBP formation likely arises from hydroxylation of aromatics or transformation of less reactive hydrophobic organic matter by hydroxyl radicals into more reactive hydrophilic organic matter (Dotson et al. 2010). This process has been shown to occur with both UV and ozone AOTs. Although increases in the regulated DBP formation potential can occur, the resulting levels may not be above the regulatory limits if the TOC is low or free chlorine contact time is minimized. Increases in DBP formation potential can be limited by evaluating the disinfection approach to limit free chlorine contact time or by evaluating alternative quenching methods, such as using granular activated carbon, which can reduce DBP precursor concentrations.

Nitrite

Nitrite formation is specific to MP UV systems. Nitrate has a high UV absorbance at the lower wavelengths emitted by MP UV lamps (i.e., <240 nm), and can be transformed (i.e., photolyzed) into nitrite and other radical species when it absorbs UV light (Mack and Bolton 1999). UV AOT systems using LPHO UV lamps will not stimulate this conversion because they do emit light below 240 nm. Sharpless and Linden (2001) found that even at nitrate concentrations of 10 mg/L (the maximum contaminant level [MCL] for nitrate), less than 1 mg/L (the MCL for nitrite in the United States) of nitrite would be formed. Nitrite formation appears not to be a significant issue, even for MP UV lamps, at least in the United States. However, in Europe the MCL for nitrite is 0.1 mg/L as N. Nitrite formation will also contribute to a higher chlorine demand. Nitrite reacts with chlorine at a ratio of 5 mg/L Cl_2 per mg/L NO_2. The increased chlorine demand can affect peroxide quenching or residual maintenance.

Bromate

Brominated by-products can also be formed when ozone and bromide are present, which must be considered for ozone AOTs because some brominated compounds are regulated by the USEPA. The most common brominated by-product associated with ozone treatment is bromate, which has an MCL under the Stage 2 D/DBPR of 10 μg/L. Bromate formation at operating ozone plants has been shown to exceed the MCL of 10 μg/L even at seemingly small concentrations of bromide in the source water (>0.2 mg/L) (Krasner et al. 1993).

With ozone AOTs, bromate formation may be controlled through the sequence of hydrogen peroxide addition (i.e., adding hydrogen peroxide prior to ozone) and/or by increasing the ratio of peroxide to ozone. Ozone reacts with the peroxide and quenches the ozone residual, thus minimizing the reaction with bromide. However, Song et al. (1997) noted that peroxide addition both increased and decreased bromate formation indicating that bromate control with hydrogen peroxide will be dependent on the water quality and operating conditions. Note that if peroxide is added upstream of ozone, disinfection credits cannot be achieved.

Bromate formation is also influenced by the pH of the water. As with carbonates, pH drives the speciation of hypobromous acid and hypobromite. The pK_a for hypobromous acid is 8.7. Hypobromite is generally considered to result in higher formation of bromate (Amy et al. 1997). Higher pHs typically result in increased formation of bromate (Krasner et al. 1993, Song et al. 1997).

The formation of bromate also needs to be considered for UV/chlorine. Recent literature has shown that bromate can be slowly formed through a series of oxidation-reduction reactions between background bromide ions and hypochlorite (Huang et al. 2008, Wang et al. 2015). Bromate may be formed through the following mechanism:

$$HOCl + Br^- \rightarrow HOBr + Cl^-$$

$$2\ HOCl + HOBr \rightarrow BrO_3^- + 3\ H^+ + 2\ Cl^-$$

This reaction was shown to proceed in the dark at a slow rate of formation. The rate of bromate formation was accelerated when UV irradiation was added. UV irradiation increased the rate of chlorine decay and the rate of bromate formation. A lower pH favored bromate formation during the studies. In general, increasing UV irradiation, higher chlorine residual concentrations, and lower pH values all increased bromate formation during UV/chlorine treatment (Huang et al. 2008). Note that bromate can also be a found in sodium hypochlorite as a by-product of the manufacturing process. Bromate formation with UV/chlorine should also be considered in the context of the potential contribution of bromate from sodium hypochlorite.

Unregulated By-products

In addition to the known regulated by-products, AOTs can form unregulated by-products that should be considered when selecting an AOT technology.

Biodegradable Compounds

Assimilable organic carbon (AOC) or biodegradable organic carbon (BDOC) are measures of the potential for bacterial regrowth or the ability of a water to support bacterial growth. Increases in AOC/BDOC may result in increases in biological growth within the distribution system. AOC can be measured using Standard Method 9217, while currently there is no standard method for BDOC analysis. Increases in AOC have been well documented with both ozone and ozone/H_2O_2 processes (Ferguson et al. 1991, Karimi et al. 1997). Information on AOC/BDOC is more limited for UV AOTs. Some testing has shown small increases in BDOC following UV/H_2O_2 treatment, while other testing has shown minimal to no increase in AOC formation after UV/H_2O_2 treatment (Swaim et al. 2007, Collins et al. 2008, Linden et al. 2015). Lui et al. (2002) showed that UV irradiation resulted in the formation of aldehydes and carboxylic acids at doses that would be typical for AOT applications (i.e., >500 mJ/cm^2). These compounds were not measured directly as AOC or BDOC but would affect the biostability of the water. AOC/BDOC formation with UV AOTs is water quality and NOM specific and should be evaluated on a case-by-case basis.

Chloropicrin and Trichloropropanone

Chloropicrin and trichloropropanone (TCP) are two potential by-products that have been linked to MP UV lamps (Reckhow et al. 2010). Formation of these compounds was not found to occur with LPHO UV lamps. Chloropicrin formation was found to be a function of both nitrate concentrations and UV dose. Thus, chloropicrin formation could be found in many MP UV AOT applications. TCP formation was found to be temperature specific with lower concentrations at higher temperature due to increasing degradation rates.

Chlorate

Chlorate formation is a potential by-product associated with UV/chlorine and has been demonstrated at the bench and full scale (Wang et al. 2015, Wetterau et al. 2015). Wang et al. (2015) demonstrated approximately 2–17 percent of photolyzed chlorine was converted to chlorate. Chlorate is not currently regulated under the Safe Drinking Water Act; however, many states have drinking water advisory levels and chlorate was included in the list of compounds for the third Unregulated Contaminant Monitoring Rule (UCMR 3).

Toxicity

Advanced oxidation technologies destroy contaminants through a transformative process that results in the molecular structure of the contaminant being altered or broken down. The breakdown products could be equally or more toxic than the parent compound. At this time, there is no standard method available for measuring toxicity, and the methods available are not a direct comparison to human health effects. Linden et al. (2015) evaluated the effects of AOTs on toxicity after treatment focusing on compounds on the USEPA's Contaminant Candidate List 3 (CCL3). In general, most of the evaluated contaminants showed no increase in toxicity based on the methods used in the study, but a select number of contaminants resulted in increased toxicity. Increases in toxicity were minimized when the AOT was followed by biological filtration. Toxicity increases have also been observed when MP UV light is used in the presence of nitrate (Martijn et al. 2015). Toxicity concerns will be site specific based on the compounds present and operating conditions. Toxicity impacts can be evaluated at the bench scale, if toxicity is of concern.

REFERENCES

Amy, G., P. Westerhoff, R. Minear, and R. Song. 1997. *Formation and Control of Brominated Ozone By-Products*. Denver, CO: AWWA Research Foundation and American Water Works Association.

Andrews. S. 2009. Canadian AOP Research/Practice for Treatment of Taste and Odour and EDC/PPCPs. *Americana*. March 17.

APHA, AWWA, and WEF. 2012. *Standard Methods for the Examination of Water and Wastewater*, Ed. Washington, DC: APHA, AWWA, and WEF.

Bolton, J.R., M.I. Stefan, R.S. Cushing, and F. Mackey. 2001. The Importance of Water Absorbance/Transmittance on the Efficiency of Ultraviolet Disinfection Reactors. In *Proc. First International Congress on Ultraviolet Technologies, June 2001, Washington, DC*. Florence, KY: CD/ROM published by the International Ultraviolet Association.

Collins, J., C. Cotton, B. Heeke, D. Dahl, and A. Royce. 2008. UV Advanced Oxidation Process for Taste and Odor Treatment: Evaluation of Assimilable Organic Formation Potential at an Indiana WTP. In *Proc. Water Quality Technology Conference, Cincinnati, OH*. Denver, CO: American Water Works Association.

Derrick, B., and E.R. Blatchley III. 2005. Field Investigations of Inorganic Fouling of UV Systems in Groundwater Applications. In *Proc. 3rd International Congress on Ultraviolet Technologies, Whistler, BC, Canada*. Florence, KY: International Ultraviolet Association.

Donhame, J.F., E.J. Rosenfeldt, and K.R. Wigginton. 2014. Photometric Hydroxyl Radical Scavenging Analysis of Standard Natural Organic Matter Isolates. *Environ. Sci.: Processes Impacts*, 16:764–769.

Dotson, A.D., V.S. Keen, D. Metz, and K.G. Linden. 2010. UV/H_2O_2 Treatment of Drinking Water Increases Post-Chlorination DBP Formation. *Wat. Res.* 44:3707–3713.

Feng, Y., D.W. Smith, and J.R. Bolton. 2007. Photolysis of Chlorine Species (HOCl and OCl⁻) with 254 nm Ultraviolet Light. *Jour. Environ. Engr. Sci.*, 6:277–284.

Ferguson, D.W., J.T. Gramith, and M.J. McGuire. 1991. Applying Ozone for Organics Control and Disinfection: A Utility Perspective. *Jour. AWWA*, 83(5):32–39.

Huang, X., N. Gao, and Y. Deng. 2008. Bromate Ion Formation in Dark Chlorination and Ultraviolet/Chlorination Processes for Bromide-Containing Water. *Jour. Environ. Sci.*, 20:246–251.

Karimi, A.A., J.A. Redman, W.H. Glaze, and G.F. Stolarik. 1997. Evaluating an AOP for TCE and PCE Removal. *Jour. AWWA*, 89(8):41–53.

Kashinkunti, R.D., K.G. Linden, G.-A. Shin, D.H. Metz, M.D. Sobsey, M.C. Moran, and A.M. Samuelson. 2003. Investigating Multi–Barrier Inactivation for Cincinnati: UV, Byproducts, and Biostability. *Jour. AWWA*, 96(6):114–127.

Keen, O.S., G. McKay, S.P. Mezyk, K.G. Linden, and F.L. Rosario-Ortiz. 2014. Identifying the Factors That Influence the Reactivity of Effluent Organic Matter With Hydroxyl Radicals. *Wat. Res.*, 50:408–419.

Krasner, S.W., W.H. Glaze, H.S. Weinberg, P.A. Daniel, and I.N. Najm. 1993. Formation and Control of Bromate During Ozonation of Waters Containing Bromide. *Jour. AWWA*, 85(1):73–81.

Kwon, M., S. Kim, Y. Yoon, T.-M. Hwang, and J.-W. Kang. 2014a. Determination of Efficient Operating Conditions for UV/Cl_2 Process. In *Proc. IUVA Americas Regional Conference*. Florence, KY: International Ultraviolet Association.

Kwon, M., S. Kim, Y. Yoon, Y. Jung, T.-M. Hwang, and J.-W. Kang. 2014b. Prediction of the Removal Efficiency of Pharmaceuticals by a Rapid Spectrophotometric Method Using Rhodamine B in the UV/H_2O_2 Process. *Chem. Eng. J.*, 236:438–447.

Lee, C., W. Choi, and J. Yoon. 2005. UV Photolytic Mechanism of *N*-Nitrosodimethylamine in Water: Roles of Dissolved Oxygen and Solution pH. *Environ. Sci. Technol.*, 39(24):9702–9709

Lin, L.-S., C.T. Johnston, and E.R. Blatchley III. 1999. Inorganic Fouling at Quartz: Water Interfaces in Ultraviolet Photoreactors I: Chemical Characterization. *Wat. Res.*, 33(15):3321–3329.

Liu, W., S.A. Andrews, J.R. Bolton, K.G. Linden, C. Sharpless, and M. Stefan. 2002. Comparison of Disinfection Byproduct (DBP) Formation From Different Water-treatment Technologies at Bench Scale. *Wat. Sci. Technol.—Wat. Supply*, 2(5-6):515–521.

Liu, W., S.A. Andrews, M.I. Stefan, and J.R. Bolton. 2003. Optimal Methods for Quenching H_2O_2 Residuals Prior to UFC Testing. *Wat. Res.*, 37(15):3697–3703.

Linden, K., U. Gunten, H. Mestankova, and A. A. Parker. 2015. Advanced Oxidation and Transformation of Organic Contaminants. Denver, CO: Water Research Foundation.

Mack, J., and J.R. Bolton. 1999. Photochemistry of Nitrite and Nitrate in Aqueous Solution: A Review. *Jour. Photochem. Photobiol. A. Chem.*, 128:1–13.

Malley, Jr., J.P., J.P. Shaw, and J.R. Ropp. 1996. *Evaluation of By-products Produced by Treatment of Groundwaters with Ultraviolet Irradiation*. Denver, CO: American Water Works Research Foundation and American Water Works Association.

Martijn, B., J.C. Kruithof, R.M. Hughes, R.A. Mastan, A.R. Van Rompay, and J.P. Malley Jr. 2015. Induced Genotoxicity in Nitrate-Rich Water Treated With Medium-Pressure Ultraviolet Processes. *Jour. AWWA*, 107(6):301–312.

Reckhow, D.A., K.G. Linden, J. Kim, H. Shemer, and G. Makdissy. 2010. Effect of UV Treatment on DBP Formation. *Jour. AWWA*, 102(6):100–113.

Rosenfeldt, E.J. 2011. Using Rapid Background Hydroxyl Radical Scavenging Measurement to Understand OH radical Reactivity of NOM in AOPs. In *Proc. Fourth IWA Specialty Conference: NOM: From Source to Tap, Costa Mesa, CA*. London: International Water Association.

Rosenfeldt, E.J., and K.G. Linden. 2007. The $R_{OH,UV}$ Concept to Characterize and the Model UV/H_2O_2 Process in Natural Waters. *Environ. Sci. Technol.*, 41(7):2548–2553.

Sharpless, C.M. and K.G. Linden. 2001. UV Photolysis of Nitrate: Effects of Natural Organic Matter and Dissolved Inorganic Carbon and Implications for UV Water Disinfection. *Environ. Sci. Technol.*, 35(14):2949–2955.

Song, R., P. Westerhoff, R. Minear, and G. Amy. 1997. Bromate Minimization During Ozonation. *Jour. AWWA*, 89(6):69–78.

Swaim, P., A. Royce, T. Smiths, T. Maloney, D. Ehlen, and B. Carter. 2007. Effectiveness of UV Advanced Oxidation for Destruction of Micro-Pollutants. In *Proc. 4th IUVA World Congress, Los Angeles*. Florence, KY: International Ultraviolet Association.

Toivanen, E. 2000. Experiences With UV Disinfection in Helsinki Water. *IUVA News*, 2(6):4–8.

Venkatesan, N., G. Hua, D.A. Recknow, and K. Kjartanson. 2003. Impact of UV Disinfection on DBP Formation From Subsequent Chlorination. In *Proc. Water Quality Technology Conference, Philadelphia, PA*. Denver, CO: American Water Works Association.

Wait, I.W, C.T. Johnston, A.P. Schwab, and E.R. Blatchley III. 2005. The Influence of Oxidation Reduction Potential on Inorganic Fouling of Quartz Surfaces in UV Disinfection Systems. In *Proc. Water Quality Technology Conference, Quebec City, QC, Canada*. Denver, CO: American Water Works Association.

Wang, D., J.R. Bolton, S.A. Andrews, R. Hofmann. 2015. Formation of Disinfection By-Products in the Ultraviolet/Chlorine Advanced Oxidation Process. *Sci. Tot. Env.*, 518–519:49–57.

Wetterau, G., P. Fu, C. Chang, and B. Chalmers. 2015. Full-Scale Testing of Alternative UV Advanced Oxidation Processes for the Vander Lans Water Treatment Facility. In *Proc. WateReuse California Annual Conference, Los Angeles, CA*. Alexandria, VA: WateReuse Association.

Zheng, M., S.A. Andrews, and J.R. Bolton. 1999a. Impacts of Medium-Pressure UV on THM and HAA Formation in Pre-UV Chlorinated Drinking Water. In *Proc. Water Quality Technology Conference, Tampa, FL*. Denver, CO: American Water Works Association.

Zheng, M., S.A. Andrews, and J.R. Bolton, 1999b. Impacts of Medium-Pressure UV and UV/H_2O_2 Treatments on Disinfection Byproduct Formation. In *Proc. AWWA Annual Conference, Chicago, IL*. Denver, CO: American Water Works Association.

7

Potential Locations for AOT Facilities

Many water utilities are interested in adding AOTs because of their ability to remove a broad range of contaminants, many of which can be hard to remove with other technologies. AOTs, depending on the design and treatment goals, can also have the added benefit of providing oxidation and disinfection simultaneously in both drinking water and reuse applications. This chapter describes potential AOT applications and locations.

SURFACE WATER APPLICATIONS

The following sections summarize considerations when locating AOT facilities for filtered and unfiltered surface water applications.

Filtered Applications

When evaluating AOTs for a filtered water application, the key decision is often whether to locate the facility upstream or downstream of filtration. Locating the AOT facility downstream of filtration often results in the best water quality within the plant (e.g., highest UVT, lowest TOC) and potentially the lowest capital and annual O&M costs for the AOT equipment (not including quenching). Water quality is typically highest postfiltration, but pre- and postfiltration chemical usage (e.g., prefiltration chlorination) should be evaluated for potential impacts on AOT sizing and operation. Locating AOTs downstream of filtration has potential drawbacks. For AOTs using hydrogen peroxide, the peroxide residual must be quenched to maintain a secondary disinfectant residual. This could include chemical quenching (e.g., sodium hypochlorite) or catalytic quenching with GAC. If GAC quenching is selected, a new GAC facility would have to be constructed (versus upgrading the existing filtration facility). AOTs can also affect the biostability of the water (i.e., increased AOC or BDOC). Locating AOTs downstream of filtration limits the ability to reduce AOC/BDOC concentrations prior to the distribution system unless GAC is used for quenching. UV/chlorine AOTs have the advantage of the residual oxidant being chlorine, which is typically used as the secondary disinfectant and would not have to be quenched.

The hydraulic grade line is also an important factor in the determining the feasible locations for an AOT system. All AOT systems will increase head loss and reduce the hydraulic grade line, which could impact both upstream and downstream processes. UV reactors alone typically have between 6 and 30 in. of head loss, which does not include the head loss for additional piping, valves, and other appurtenance. Ozone systems could have between 6 and 60 in. of head loss depending on the design of the contactor (i.e., tank versus pipeline). Quenching, if required, can also add additional head loss depending on the selected approach. Available head and potential operational changes should be considered when siting an AOT facility.

Many filtered water utilities are evaluating the application of UV disinfection for meeting regulatory requirements or as an additional pathogen barrier. When evaluating UV for disinfection purposes, some utilities are also evaluating the potential for including advanced oxidation. UV AOTs have the benefit of providing simultaneous disinfection and advanced oxidation. These dual-purpose systems can also have the advantage of turning lamps or reactors off to minimize operating costs, if advanced oxidation is only required seasonally (e.g., taste-and-odor events). The USEPA regulations for UV disinfection credit focus on postfilter applications (USEPA 2006). Therefore, for applications considering disinfection and advanced oxidation, the UV reactors should be located downstream of filtration. If the AOT facility is desired prior to filtration, the utility would need to coordinate with its primacy agency to determine the requirements for UV disinfection credits.

Ozone AOTs can also be used for disinfection credit, but this may limit the sequence of ozone and hydrogen peroxide application. For disinfection credit, ozone would be added first and hydrogen peroxide would be added after CT credit is achieved. Most ozone applications are prefiltration with the filtration process being converted to biofiltration to limit AOC/BDOC concentrations in the finished water.

Along with water quality considerations, the location of the AOT facility should also take into account the proximity to high service pumps that may create water hammer issues. Water hammer can result in lamp break events with UV AOT systems. A surge analysis may be necessary to verify that water hammer will not be an issue. A hydropneumatic tank or pressure-relief values may be necessary if water hammer is a concern.

The utility should evaluate the advantages and disadvantages for both pre- and postfiltration locations, which are summarized in Table 7-1. To select the most appropriate location, design criteria, hydraulic constraints, disinfection requirements, footprint, and overall cost will need to be evaluated at each potential location.

Unfiltered Applications

UV and ozone AOTs can provide similar benefits on unfiltered water systems for contaminant oxidation and the potential for disinfection. Locating the AOT facility for an unfiltered application focuses more on the source of the target contaminants. Many unfiltered systems have a covered or uncovered storage reservoir included as part of the

Table 7-1   Filtered surface water AOT locations

Location	Advantages	Disadvantages
Upstream of filtration	• Potential for biofiltration for AOC/DBOC removal with existing filters • Potential for quenching through filtration	• Potentially lower UVT, higher scavenging demand, and TOC leading to increased capital and annual O&M costs • Additional coordination for disinfection credit with UV AOTs
Downstream of filtration	• Typically best water quality for reduced capital and annual O&M costs • Potential for disinfection credit with UV AOTs	• Additional facilities may be required for quenching or AOC/DBOC removal

water system, which can be a source of taste-and-odor concerns. For uncovered reservoirs, the AOT facility would typically be located downstream of the reservoir to provide disinfection credits and treatment of any target contaminants that may be present in the reservoir. If the source of the contaminant is upstream of a covered reservoir, the facility could be located on the influent or discharge of the reservoir/tank. The most appropriate location depends on hydraulic and space constraints and the source of the contamination. Water quality issues, such as the following issues, may also help determine whether the AOT facility is most appropriate before or after the storage reservoir/tank:

- Changes in water quality caused by reservoir turnover
- Algae counts in the reservoir
- Location of treatment chemicals
- Water quality variability prior to and after the reservoir

Important water quality issues are described in more detail in chapter 6.

GROUNDWATER APPLICATIONS

Groundwater systems may consider installing AOTs for a range of contaminants prior to the distribution system for drinking water or for remediation of contaminated groundwater. 1,4-dioxane and *N*-nitrosodimethylamine (NDMA) are two examples of groundwater contaminants that are best treated with AOTs. AOTs are also highly effective at treating many volatile organic compounds (VOCs). However, AOTs are not currently considered a best available technology (BAT) for VOC treatment.

For groundwater systems, the AOT facility may either be installed at individual wellheads or at a centralized facility if multiple contaminated wells are located in the same area. For systems with multiple wells, centralized versus decentralized treatment can be evaluated. Water quality and contaminant concentrations can vary between wells and should be considered. Individual wellhead treatment may be advantageous if it allows for treatment of a higher concentration of contaminants at a lower flow rate. As with an AOT facility at a WTP, an engineering cost analysis can be performed to determine the most appropriate location for a groundwater AOT facility. Groundwater applications should also be evaluated to determine if water hammer is an issue, considering the proximity of the AOT equipment to the wellheads. Head loss through the AOT system should also be carefully considered as the additional head loss from an AOT facility will decrease the production capacity of existing wells.

REUSE APPLICATIONS

As surface water and clean groundwater resources are becoming scarce around the country, many utilities are looking toward wastewater recycling or reuse. The reuse water can be used for potable water augmentation (i.e., direct or indirect) and/or nonpotable reuse. For many indirect and direct potable reuse applications, AOTs are being considered for their ability to remove a broad range of contaminants that are difficult to remove with other technologies (e.g., NDMA and pharmaceuticals and personal care products). Many of these contaminants can make it through other advanced treatment technologies, such as reverse osmosis. AOTs can also provide high levels of virus inactivation, which is often an important design parameter for reuse applications.

UV AOTs are typically most cost-effective if they are applied to water with the highest UVT and lowest scavenging demand. In most reuse applications, this follows membrane filtration. After nanofiltration or reverse osmosis, the UVT of the water is often 95 percent or higher, which increases the efficiency of UV AOTs. Postmembrane filtration, the TOC and alkalinity are also typically low, which reduces the hydroxyl radical scavenging demand of the water. However, the hydroxyl radical scavenging demand could be affected if the membrane permeate contains a chloramine residual. Chloramines can decrease the UVT of the water and can also impart a hydroxyl radical scavenging demand. The influence of a chloramine residual will also be a function of the concentration and should be considered when sizing an AOT system.

Postmembrane filtration locations are also favorable for UV/chlorine applications. The efficiency of a UV/chlorine process is highly pH dependent and is more efficient at a pH at or below 6, which is typical of reverse osmosis permeate. However, breakpoint chlorination will be required to achieve a free chlorine residual if chloramines are present.

Ozone AOTs have also been gaining interest in reuse applications. Ozone AOTs are not as effective for NDMA treatment and have been shown to increase NDMA formation. However, ozone can be a cost-effective alternative for a broad range of contaminants. Ozone has also been shown to reduce biological fouling on membranes (Stanford et al. 2013, Serna et al. 2014). For this reason, ozone AOTs are mainly located upstream of membrane filtration. Membrane filtration also helps to reduce bromate concentrations that can be a by-product of the ozone AOTs.

REFERENCES

Serna, M., R.S. Trussell, and F.W. Gerringer. 2014. *Ozone Pretreatment of a Non-Nitrified Secondary Effluent Before Microfiltration*. Alexandria, VA: WaterReuse Research Foundation.

Stanford, B., A.N. Pisarenko, S. Snyder, and R.D. Holbrook. 2013. *Pilot-Scale Oxidative Technologies for Reducing Fouling Potential in Water Reuse and Drinking Water Membranes*. Alexandria, VA: WateReuse Research Foundation.

USEPA. 2006. *Ultraviolet Disinfection Guidance Manual for the Final Long Term 2 Enhanced Surface Water Treatment Rule*. Washington, DC: US Environmental Protection Agency, Office of Safe Water. http://www.epa.gov/safewater/disinfection/lt2/pdfs/guide_lt2_uvguidance.pdf.

8

AOT System Design Considerations[1]

This chapter describes key elements to consider when planning and designing an AOT system. This chapter is not intended to provide an exhaustive list of design considerations but rather provides discussions of key elements that should be considered when designing an AOT facility.

TREATMENT GOALS

The specific objectives of an AOT system should be clearly defined before the design begins. The goals for the facility could be regulatory compliance, aesthetics (e.g., taste and odor), or public health protection for unregulated contaminants. Understanding the specific goals of the project helps to ensure that the design meets the utility's and the regulatory agency's expectations based on the potential regulatory requirements, target contaminants, source water quality, and the overall treatment strategy.

The regulatory requirements for AOT facilities vary depending on the application, source water, target contaminant, and current water treatment process. Discussions with the regulatory agency should be initiated at the start of the project to determine any specific regulatory and design requirements.

If the treatment goals of the AOT system include disinfection, the ozone and UV equipment design should comply with requirements and regulations for those individual technologies and purpose. This handbook does not describe the steps required for disinfection credit, and other USEPA documents should be referenced for disinfection system design (see also Bolton and Cotton 2008).

KEY DESIGN CRITERIA

Several key design criteria common to UV and ozone AOT systems require consideration during the planning and design phase.

Target Contaminant and Destruction

The AOT design depends on the target contaminant and the design influent and treated water concentrations. Disinfection applications generally have a limited number of

---

1 Some of this chapter has been adapted (with permission) from Bolton and Cotton (2008).

well-defined target pathogens (i.e., *Giardia, Cryptosporidium,* or virus). AOT applications have the potential for a much broader set of target contaminants that are both regulated and unregulated. The effectiveness and corresponding system sizing are directly related to the hydroxyl radical rate constants for the target contaminant (appendix B) (except for UV photolysis application). Contaminants with higher rate constants result in smaller less expensive AOT systems. Thus, identifying the target contaminant is critical for system sizing. As with other water quality parameters, historical data should be reviewed to determine design influent concentrations, and if historical data are limited, additional monitoring is recommended. Consideration should be given to seasonal variations, increasing or decreasing trends, and the contaminant source. For example, AOT systems treating contaminated groundwater plumes should consider concentrations throughout the plume and potential movements of the plume relative to the production wells.

In addition to understanding the target contaminants influent concentrations, understanding the treated water concentration and corresponding log reduction or percent reduction is equally important. The target treated water concentration should consider regulatory requirements, if applicable, and may consider current method detection limits for the target contaminants. Equipment sizing is proportional to the necessary reduction. Overly conservative influent and/or treated water goals can result in a larger than necessary facility or increased annual O&M costs arising from inadequate turndown capacity.

For some applications, such as reuse applications, a surrogate is often used to cover a range of contaminants (e.g., 0.5-log 1,4-dioxane reduction). Sizing is based on the surrogate but treatment will result in the destruction of a range of contaminants.

Flow Rate

Flow rate is a key design consideration as it can dictate the size of the system and chemical usage. Many UV and proprietary ozone systems have specific flow rate limitations, which may determine the overall system size and footprint depending on the water quality.

Generally, the key flow-rate design parameter for an AOT system is the maximum flow rate. Identifying the average and minimum flow rate is also necessary to optimize operating costs as well as allowing proper turndown capacity to cover the full range of operating conditions. The maximum flow rate is also important if the system is designed for disinfection. UV reactors have a maximum validated flow rate under which they can operate, and flow rates higher than design could limit the disinfection credit. Note that UV reactors are regulated based on the maximum instantaneous flow rate and not a daily average flow rate. Similarly, flow rates higher than design can limit the ability of an ozone AOT to meet the required CT as CT must be achieved prior to addition of hydrogen peroxide.

For seasonal applications, the flow rate should be evaluated during the time of year that treatment is expected to be required. If the maximum flow rate does not occur during the targeted treatment season, capital savings could be possible by reducing the equipment size and building footprint.

Hydroxyl Radical Scavenging Demand

The majority of AOT systems rely on hydroxyl radicals for contaminant destruction. Hydroxyl radicals are nonselective and oxidize compounds other than the target contaminant. The hydroxyl radical scavenging demand must be properly understood and accounted for in the design. Water quality parameters that affect the hydroxyl radical scavenging demand (e.g., alkalinity, NOM, pH) should be analyzed including any potential seasonal variations. Some AOT vendors also request water samples to experimentally measure the hydroxyl radical scavenging demand. Any water samples collected for designing the AOT system should be representative of the expected water quality and collected at the expected location (e.g., pre- or postfiltration) for the AOT facility.

By-product Precursors

As oxidative technologies, all AOTs have the potential to create by-products from the breakdown of the target contaminant or reactions with other compounds within the water. Water quality parameters that could affect by-product formation include

- Nitrate (MP UV AOTs only)
- Bromide (ozone and UV/chlorine only)
- Organic matter
- pH

These parameters should be evaluated and provided to the AOT vendors for consideration during design.

Downstream biological filtration can help reduce the potential increase in DBP formation potential and increase the biological stability of the water. Biological filtration is often recommended and may be required in some countries following ozonation for reducing AOC/BDOC concentrations prior to the distribution system. Guidance for implementing and designing biological filtration can be found in several resources (Randtke and Horsley 2012, Lauderdale et al. 2011).

Turndown Capacity

AOT systems are often designed to treat the worst-case contaminant concentrations at the highest flow and worst water quality. During operation, it is unlikely or infrequent that all three worst-case parameters will occur simultaneously. AOT systems typically have high annual O&M costs due to power and chemical requirements. Optimizing treatment is important for minimizing costs. Turndown capacity is important when evaluating the design to provide the ability to adapt to actual operating conditions. Several factors should be considered when evaluating turndown capacity including

- Expected range of normal and worst-case operating conditions for contaminant concentrations, flow, and water quality.
- Number of treatment trains to adjust to flow rate changes.
- Online instrumentation or treatment target inputs to allow reduced treatment levels when appropriate.
- Ability to turn down UV lamps or turn lamps/reactors off for UV AOTs.
- Ozone generator turndown for dose pacing at lower flows.
- Ozone gas flow control valve turndown and metering accuracy at lower flows.
- Number of chemical metering pumps and turndown capacity for dose pacing based on expected range of flows and treatment targets.

Redundancy

Redundancy is an important design consideration for all AOT applications. For most disinfection applications, N+1 is a standard minimum redundancy requirement. However, AOT systems may not include a fully redundant UV reactor or ozone generator. AOT equipment is expensive and oftentimes may treat contaminants that are unregulated. Because of the higher dose requirements for AOT systems, many facilities may include multiple treatment trains. Having multiple treatment trains may provide inherent redundancy at operating conditions that are more favorable than the design conditions. To optimize capital and annual O&M costs, some utilities may choose to not include redundant equipment at the design conditions for worst-case contaminant levels, flow rate, and water quality. Considerations for redundancy include

- Regulatory requirements.
- Ability to reduce facility flow rate or to shut down the facility during equipment maintenance.
- Excess equipment available under normal operating conditions.
- Number of UV reactors or ozone equipment: if only one reactor or generator is required, redundancy may be desired.

Redundancy is generally recommended for all chemical feed pumps. Chemical metering pumps are relatively inexpensive and prone to failure, especially with chemicals prone to off-gassing (e.g., hydrogen peroxide and sodium hypochlorite). Chemical metering pumps are typically designed with an N+1 configuration.

UV AOT DESIGN CRITERIA

The key design criteria specific to UV AOT systems include the UV transmittance (UVT) of the water and fouling–aging factor.

UVT

UVT is an important water quality parameter for designing a UV AOT system. The UV system size depends on the UVT, and selection of the design UVT is important for both the UV equipment selection and capital and operating costs. Overly conservative design UVT values (i.e., low UVT) can result in the need for additional equipment, which increases the overall facility cost. If the equipment is overdesigned and does not have the proper turndown capability, overdosing may occur during normal operation, which results in increased operating costs. However, if the design UVT is inappropriately high, the UV system may be unable to provide an adequate amount of UV energy to form the target concentration of hydroxyl radicals during low UVT events.

Available UVT data should be evaluated to define both the design UVT and the range of UVTs that can be expected. The UVT data should be also evaluated in conjunction with flow rate data, if available. The design UVT and the minimum UVT may not be the same value, depending on the flow and UVT data. The design UVT may be higher than the minimum UVT if the low UVT events occur simultaneously with low water demand. This may be the case for some seasonal AOT applications (e.g., taste-and-odor compounds), where the UVTs may be higher during the summer months.

Understanding the range of UVTs can also help with equipment selection. If the design UVT and average UVT are significantly different, it may be advantageous to select UV equipment with high turndown capabilities or the ability to turn lamps on and off to optimize power use during higher UVT events.

UVT data are typically reported at a wavelength of 254 nm. If MP UV reactors are being considered, measuring the UVT spectrum of the water between 200 to 300 nm is also important as hydrogen peroxide and chlorine both absorb light outside of 254 nm. The UV absorbance typically increases (i.e., decreasing UVT) at wavelengths below 254 nm in natural waters. The increased UV absorbance and decreased UVT at the lower wavelengths can affect the generation of hydroxyl radicals with MP UV reactors. Compounds such as nitrate can have a high UV absorbance below 254 nm. Any seasonality in nitrate should be considered when evaluating the full UVT spectrum.

AOT systems can be installed in several locations within a WTP. The UVT of the water may change through the treatment process and UVT measurements should be collected from the potential locations, if possible. If UVT data are not available, sampling is recommended, but the duration of the sampling period and frequency of sample collection depend on the variability of the source water quality. For example, a water utility with very stable TOC concentrations (e.g., typical groundwater wells) is likely to have stable UVT measurements and may need only 1 or 2 months of weekly samples. A water utility that experiences seasonal changes, however, would benefit from more frequent data collection during seasonal events and over a longer period (6 to 12 months or more). As described previously, samples should not be filtered or pH-adjusted before measuring the UVT to avoid biasing the UVT high.

Fouling–Aging Factor

The performance of a UV reactor can be affected by the fouling of the quartz sleeves and UV intensity sensor windows and by lamp and sleeve aging. These factors should be accounted for in the facility design to minimize the potential for reduced treatment capacity.

Sleeve and Sensor Window Fouling

Fouling is typically caused by precipitation of compounds with low solubility, compounds for which the solubility decreases as temperature or oxidation-reduction potential (ORP) increases (e.g., iron compounds), or by photochemical reactions. Fouling rates are dependent on the site-specific water quality and can depend on ORP, hardness, alkalinity, lamp temperature, pH, and the presence of certain inorganic constituents (e.g., iron, manganese, and calcium) (USEPA 2006).

These water quality parameters are helpful for the UV manufacturers to qualitatively assess the fouling potential for their UV reactors and to determine the need for specific cleaning systems. If the data are not available, sampling is recommended. The USEPA's *UV Disinfection Guidance Manual* (UVDGM) (USEPA 2006) provides guidance on the recommended frequency of sampling. For AOT systems, chemical oxidants (e.g., chlorine) are added upstream of the UV reactor, which can increase the fouling potential. Fouling has typically been low for reuse applications that treat RO permeate due to the low pH and aggressiveness of the water but can be present on groundwater and surface water applications. Pilot-scale or demonstration-scale testing can be performed to determine the site-specific fouling rate and cleaning efficiency, if desired.

Lamp, Sleeve, and Sensor Window Aging

UV lamps, sleeves, and sensor windows age as the UV equipment is used, which causes a reduction in available UV light over time. UV lamps have a reduction in UV light output at a rate that depends on the hours of operation, number of on/off cycles, the water temperature, power applied per unit (lamp) length, and heat transfer from lamps (USEPA 2006). Lamp aging curves are typically available from the UV manufacturer and should be verified by an independent third party. Sleeve and sensor window aging can also be caused by prolonged exposure to UV light; however, typically this effect is considered to be less of an issue than lamp aging if sleeves and sensor windows are replaced as recommended by the UV manufacturer.

Combined Fouling–Aging Factor Considerations

The fouling–aging factor is site specific and is based on an assessment of fouling and aging information specific to the site water quality and selected equipment, and typically ranges from 0.4 to 0.9 (NWRI 2012). The fouling factor represents the decrease in the fraction of UV light passing through a fouled sleeve as compared to a new sleeve (USEPA 2006). The aging factor represents the decrease in lamp output due to aging of the lamp or sleeves compared to new lamps and sleeves. The fouling–aging factor is calculated by multiplying the fouling factor and the aging factor and is used by the equipment vendor when sizing the equipment. The UV AOT facility design should specify a fouling–aging factor coupled with a guaranteed lamp life in the design criteria that are provided to the UV manufacturer.

For example, a fouling–aging factor of 0.8 may be appropriate if

1. The UV reactors being installed have automatic cleaning, and there is little fouling potential based on the water quality evaluation.

2. The lamp aging data show the lamps only lose 10 percent of their intensity over the guaranteed life.

OZONE KEY DESIGN CRITERIA

The key design criteria specific to ozone AOT systems include the water quality and downstream treatment.

Ozone Dose

The target level of treatment and source water quality determines the target ozone dose. The target ozone dose will be a function of the ozone demand and decay characteristics of the water and the location of the peroxide addition. Ozone demand is the fast initial demand for ozone, which is followed by the slower first-order decay of ozone (Figure 8-1). Ozone has a relatively short half-life in water resulting in decreasing concentrations as a function of time. Both the ozone demand and decay characteristics should be evaluated to properly determine the design ozone dose. The ozone demand and decay are a function of the water quality parameters, such as temperature, pH, organic matter, metals, alkalinity, and organic or synthetic compounds that can be oxidized.

Oxygen Feed Systems

Historically, there were many air-fed ozone generator systems constructed; however, the majority of large ozone facilities currently being installed are oxygen-fed. Oxygen can either be produced on-site using pressure swing adsorption (PSA) or vacuum pressure swing adsorption (VPSA), or it can be trucked in and delivered as liquid oxygen (LOX) for vaporization. An economic analysis is often employed to determine the most suitable system considering LOX cost in the local market versus the capital and O&M costs associated with oxygen generation on-site. When considering on-site oxygen generation, equipment maintenance and useful life must also be factored in.

Ozone Generator Alternatives

Several ozone generator alternatives are available on the market. Depending on system size, different technologies are available. The majority of the municipal market is supported by tube-type corona discharge generators; however, for smaller applications

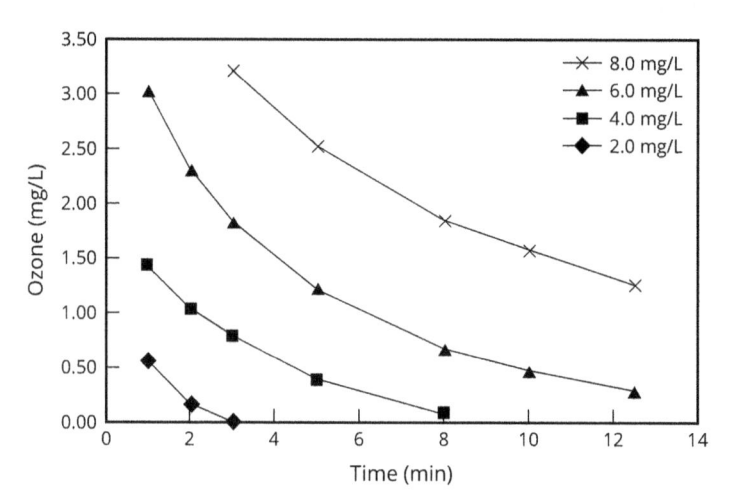

Raw water. T = 25 deg. C, pH = 7.9, TOC = 5.34 mg/L.

Figure 8-1   Example of an ozone demand and decay curve at 25°C.

modular plate-type corona discharge generators are becoming available in the market. The most appropriate technology is determined by evaluation of capital cost, generator efficiency, and specific requirements of the installation, such as space availability, ozone production requirements, and water temperatures (for cooling water).

Cooling Water Systems

A major by-product of ozone generation is heat, and heat must be continually removed from the generators to keep them operating efficiently. Cooling of ozone generators can be done in several different ways depending on cooling water availability, temperature, and quality. Once-through cooling is sometimes used, in which a water supply is provided to the generators, run through the generator, taking away heat, and discharged elsewhere. Water quality must be thoroughly understood before using once-through cooling because of the dangers of fouling or corrosion within the generator. Closed loop cooling is often used, and includes an open (once-through) loop and a closed loop with a heat exchanger. The closed loop water quality can be stabilized with chemical additives or can be deionized water to prevent corrosion or fouling within the generator. In areas where water temperatures are high, closed loop with chiller systems are sometimes used. A life-cycle cost analysis is often necessary to determine if the improvement in ozone efficiency offsets the cost of running and maintaining a chiller, but in higher temperature environments, a chiller is sometimes beneficial.

Ozone Diffusion System

In addition to the selected ozone dose, the ozone transfer efficiency must also be considered. Ozone transfer efficiencies vary depending on the selected ozone injection

Table 8-1   Ozone injection advantages and disadvantages

Sidestream Injection		Fine-Bubble Diffusion	
Advantages	Disadvantages	Advantages	Disadvantages
Higher transfer efficiency of ozone gas into water (~95%)	Additional equipment required	Simple operation, and no moving parts other than valves	Lower transfer efficiency of ozone gas into water (~90%)
Works well with lower flows	Turndown capability limited by injection system	No additional pumping energy needed for delivery of ozone	Maintenance of gaskets and piping
Does not require entry into ozone contactor for maintenance	Power required for sidestream pumping		Potential scaling of diffusers in some waters
Can be used with pipeline or shallower contactor structure	Space required for equipment		Requires isolation and entry into contactor for maintenance
			Requires +20 ft deep contactor

methods. Sidestream injection can have ozone transfer efficiencies of up to 95 percent while fine-bubble diffusers have transfer efficiencies of approximately 90 percent. The ozone generators must be sized large enough to cover the design doses while accounting for overall transfer efficiency. In addition to overall transfer efficiency, each ozone injection method has distinct advantages and disadvantages that should be considered (Table 8-1).

CHEMICAL FEED SYSTEM DESIGN

For AOTs that rely on chemical addition for radical formation (e.g., hydrogen peroxide or chlorine), the required chemical dose to achieve the necessary level of treatment must be known. For UV AOTs, treatment is achieved through a combination of chemical dose and energy input. Higher chemical doses can reduce the required energy input for the same level of treatment. Oxidants can add to the hydroxyl radical scavenging demand of the water depending on pH and concentration. As a result, there may be diminishing returns for increased treatment efficiency with higher oxidant doses. To determine the optimum combination, the cost of the chemical and added hydroxyl radical scavenging demand should be weighed against the cost of power and the capital costs for larger systems.

As with any chemical feed system design, understanding the design, average, and minimum chemical dose is important to a successful design. Typically, the design and minimum chemical doses combined with the design and minimum flow rates are used to verify that the chemical metering pumps have the proper turndown capacity to cover the range of expected operating conditions. Understanding the average flow and average chemical dose is also important for sizing the chemical storage tanks, which typically are designed for a minimum of 30 days of storage. Hydrogen peroxide does not rapidly degrade, compared to sodium hypochlorite, with typical degradation rates of less than 1 percent per year in large storage tanks at normal ambient temperatures (US Peroxide 2015). Smaller systems may consider sizing hydrogen peroxide storage tanks based on a full truck delivery instead of 30 days storage, to reduce chemical delivery costs.

Both hydrogen peroxide and sodium hypochlorite are prone to off-gassing. Off-gassing can bind some diaphragm metering pumps if they are not properly designed. Selection of diaphragm pumps with built-in venting/off-gassing capability or peristaltic metering pumps can lead to reduced pump priming issues during operation. Design considerations related to off-gassing include

- Valves should be diaphragm valves or vented ball valves to avoid trapping small amounts of liquid between the ball and the casing. The trapped liquid can cause the valve to explode if enough pressure builds up due to off-gassing.
- Maintain flooded suction for chemical metering pumps. Even though metering pumps have a manufacturer-published capability to pull a suction lift, it is not recommended in the situation of pumping a neat chemical that is prone to off-gassing.
- Minimize or avoid having any localized high points in a chemical feed line where a gas bubble could become trapped and vapor lock the pump or obstruct flow.
- Degassing valves installed at high points to relieve off-gassed vapors. These operate similar to automatic air release valves but are specifically manufactured for chemical systems, with the proper chemical resistance of all wetted parts.

The off-gassing potential of hydrogen peroxide must also be considered when designing the chemical storage. Storage of hydrogen peroxide should be designed to minimize contamination and provide proper venting. Contamination of the bulk hydrogen peroxide can lead to rapid degradation, which results in off-gassing and heat generation. The storage tank must be properly vented to account for potential off-gassing events. The feed system design should also avoid any situation where hydrogen peroxide could be contained without proper venting (e.g., piping between two valves without a degassing valve). A water supply should also be available in case of a spill to dilute the hydrogen peroxide.

As with all chemical storage and feed systems, chemical compatibility is critically important. Hydrogen peroxide is incompatible with organic materials and many metallic materials with the exception of stainless steel and aluminum. All metallic tanks, fittings, valves, and piping material in contact with concentrated hydrogen peroxide should be stainless steel or aluminum. High density polyethylene (HDPE) and polyvinyl chloride (PVC) materials can also be considered if the hydrogen peroxide concentration is less than 52 percent.

Sodium hypochlorite storage design should consider the trade-off between tank sizes that allow for adequate storage and the loss in strength that occurs over that time period. Sodium hypochlorite degradation can be addressed by locating storage tanks inside a shaded area or temperature-controlled space, or it can be addressed operationally by cycling the tank more frequently and not filling it completely.

The chemical storage and feed system should also be located as close as possible to the dosing location to avoid long lengths of chemical feed piping. As an alternative, use of a water dilution line could keep the residence time in the chemical feed line to a minimum. Doing so can help to avoid some of the issues that arise in a neat chemical feed lines, such as off-gassing, vapor purging, or draining/filling operations before and after a repair.

Chlorine is a regulated disinfectant with a USEPA maximum residual disinfectant level (MRDL) of 4 mg/L. Additionally, chlorine can cause pitting of the stainless-steel shell of a UV reactor. Chlorine concentrations should be maintained less than 5 mg/L for type 316 stainless steel to avoid pitting (US Peroxide 2015).

HYDROGEN PEROXIDE QUENCHING

AOTs that rely on hydrogen peroxide must also consider the residuals after treatment. Hydrogen peroxide must be quenched in order to maintain a free chlorine residual and can increase corrosion downstream of the AOT facility. For some indirect potable reuse applications, the hydrogen peroxide residual is not required to be quenched when the water is directly injected into a well. For drinking water applications, the hydrogen peroxide residual must be quenched. Understanding the expected residual is necessary to properly size the downstream quenching systems. The costs of quenching should also be considered when selecting the design chemical dose.

Ozone/H_2O_2 facilities typically have lower peroxide residuals because peroxide addition can be based on the reaction stoichiometry of the reaction between peroxide and ozone. However, the peroxide residuals can be higher if excess peroxide is necessary to limit bromate formation. For the UV/H_2O_2 process, excess peroxide is necessary because peroxide does not readily absorb UV light. Therefore, both ozone/H_2O_2 and UV/H_2O_2 could have a peroxide residual that requires quenching in order to have a chlorine residual in the distribution system and to maintain a nonaggressive water quality.

Quenching can be accomplished chemically or through a catalytic reaction with GAC or biologically active carbon (BAC). Commonly listed chemical quenching agents include sodium hypochlorite, sodium sulfite, and sodium or calcium thiosulfate. Most operating facilities use sodium hypochlorite because of the low stoichiometric dose required for quenching, and it is typically already on-site for disinfection. Sodium sulfite and sodium or calcium thiosulfate have been shown to have complex reactions and may require dosing in excess of the stoichiometric dose in order to achieve rapid quenching (Collins et al. 2011, Keen et al. 2013). The excess residuals add an additional chlorine demand to the water, which makes these chemical impractical for full-scale applications.

Ozone/H_2O_2 and UV/H_2O_2 can also affect the required concentrations of quenching agents because of the complex radical reactions occurring. For example, the stoichiometric requirement for sodium hypochlorite is approximately 2.09 mg/L Cl_2 for 1 mg/L H_2O_2. Actual sodium hypochlorite doses may be higher due to increased chlorine demand created by organic transformations in the AOT process or nitrite formation (Cotton et al. 2010, Dotson et al. 2010, and Venkatesan et al. 2003).

DBP formation is a concern when adding chlorine to natural waters. Despite the higher chlorine doses required to quench hydrogen peroxide, the chlorine reaction with peroxide is faster than the reaction with organic matter, which limits DBP formation during quenching (Keen et al. 2013). Total DBP formation may also be similar for the potential quenching chemicals when considering the overall contact time in the distribution system (Figure 8-2).

The high sodium hypochlorite doses required for peroxide quenching can affect other water quality parameters including increased pH, sodium, chloride, TDS, and bromate concentrations. Bromate can be formed during the sodium hypochlorite manufacturing process and high doses of sodium hypochlorite can result in measurable bromate concentrations. Bromate levels could be reduced through strict limits within the sodium hypochlorite specifications.

GAC quenching evaluations and full-scale operation have shown rapid peroxide removal rates with high loading rates and low empty bed contact times (EBCTs) on the order of approximately 5 minutes EBCT (Collins et al. 2011, Festger 2014, Collins et al. 2014). However, limited information is currently available to fully understand

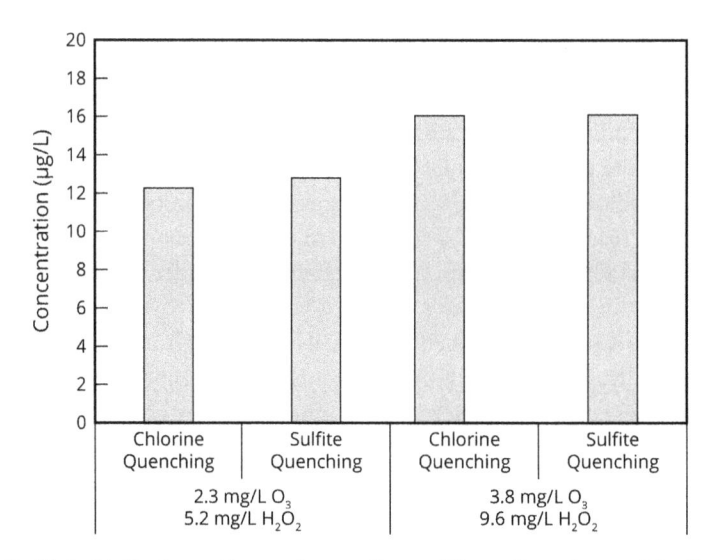

Figure 8-2 Total trihalomethane formation with sodium hypochlorite and sodium sulfite peroxide quenching (3-day simulated distribution system [SDS]. 1–1.2 mg/L Cl_2) (Adapted from Collins et al. 2011).

Table 8-2 Comparison of sodium hypochlorite and GAC for peroxide quenching

Peroxide Quenching Method	Advantages	Disadvantages
Sodium Hypochlorite	• Lower capital costs • No additional head loss	• Potentially higher annual O&M costs at high peroxide residuals • More difficult to control with fluctuations in peroxide concentrations • Can negatively impact water quality
GAC	• Potential to oxidation by-products • Stable operation due to ability to handle fluctuations in peroxide concentrations without reduced quenching performance • Potentially lower annual O&M costs • Minimal maintenance requirements (i.e., long GAC bed life and infrequent media change out)	• Higher capital costs • Additional head loss that may require booster pumping • Backwash supply and backwash waste handling may be needed

the impact of various water qualities on quenching kinetics, and quenching should be evaluated given site-specific water quality. The low EBCT requirements for peroxide quenching allow for minimizing the volume of GAC required for treatment and increasing the loading rate. This allows for fewer contactors at the full scale, which reduces capital costs and the facility footprint compared to typical GAC adsorption facilities. If GAC pressure contactors are used, contactor design should include proper air release valves and back-washing capabilities to avoid overpressurization of the GAC contactor from oxygen buildup, which is a by-product of the quenching reaction.

Chemical and GAC quenching both have distinct advantages and disadvantages that should be considered when evaluating peroxide quenching (Table 8-2).

HYDRAULICS

The hydraulic design of an AOT system can affect performance and the hydraulic profile of the treatment facilities. This section describes the hydraulic issues to be considered.

AOT System Head Loss

Head loss through an AOT system is dependent on the system design. LPHO UV reactors typically have higher head loss than MP UV reactor because of the higher number of lamps in a LPHO reactor. Some proprietary ozone AOT systems with serpentine pipe contactors with inline static mixers have a higher head loss than a more conventional ozone contactor design. If GAC quenching is used, the GAC contactors add additional head loss to the system. AOT systems can result in additional head loss that may require booster pumping or a reduction in the hydraulic grade line to compensate for the additional head loss.

Methods for Addressing Head Loss

Changes to the plant design or operation may be required if the head loss through the AOT system is greater than the available head. Potential modifications that could be considered alone, or in combination, include the following (USEPA 2006):

- Eliminating existing hydraulic inefficiencies within the facility to improve head conditions (e.g., replacing undersized or deteriorated piping and valves or upgrading flowmeters or flow control valves).
- Modifying the operation of the downstream clearwell (e.g., lowering the surface water level).
- Modifying the operation of the filters (e.g., increasing the water elevation above the filters or shortening the filter runs).
- Installing booster pumps.
- Modifying the operation of the high service pumps (HSPs) (only applies if the AOT system is close to the HSPs).
- Reducing well production capacity for groundwater applications.

Water Level and Air Release

UV AOT reactors are typically closed vessel and designed to be pressurized (i.e., flow full) to reduce the possibility of lamps overheating and for optimal hydraulics. Therefore, the facility hydraulic design needs to have the UV reactors below the downstream hydraulic grade line elevation. In addition, air release valves may be necessary in upstream or downstream piping to release air from the piping, especially during reactor filling. The valve locations are dictated by the specific configuration of the facility. UV reactor pressure limits typically range from 60 to 150 psi depending on the model. UV reactors are designed for positive pressures and have minimal ability to handle vacuums. Potential system pressures and vacuums should be considered relative to the selected equipment.

Ozone contactor design and off-gas destruct design should also account for air handling. Rapid filling of ozone contactors may be possible depending on the facility hydraulics. Conventional ozone contactors maintain a vacuum to capture and treat any ozone off-gas. The off-gas system should be designed to handle the air flow rate both during normal operation and during contactor filling.

ELECTRIC POWER SYSTEMS

Electrical issues to be considered in the design are described in this section. Issues concerning power reliability and backup power are also discussed in this chapter.

Power Supply

The electrical power configuration should take into account the power requirements of the selected AOT equipment, the reliability objectives, and power quality issues. Excluding high service pumping, the electrical load from an AOT system is typically among the larger loads at the WTP. The overall power requirements can vary depending on the selected equipment and water quality. Ozone AOTs typically have lower electrical requirements compared to UV AOTs.

The AOT system manufacturer should be consulted to determine the power supply voltage and total load requirements for the selected equipment. Some equipment can induce current and voltage harmonic distortion that can damage other electrical equipment. The overall AOT system design and equipment should meet the Institute of Electrical and Electronic Engineers (IEEE) 519 Standard to minimize the risk of harmonic distortion.

Power Quality

Both UV and ozone AOTs require reliable power quality. UV and ozone equipment can be highly sensitive to power quality events such as voltage sags/swells or power interruptions. These power quality events can result in the UV lamp losing its arc (i.e., intensity) and thus reducing or eliminating treatment. Ozone generators can also be sensitive to power quality events and can risk treatment if the generators shut down.

A power quality assessment is recommended if the site is (1) known to have power quality problems (e.g., 30 power interruptions and/or brownouts per month) or (2) located in a remote area where the power quality is unknown (USEPA 2006). For both UV and ozone AOT systems, existing power supply data may be sufficient to determine if power quality problems exist, but a power quality monitor can be installed if additional information is required. Note that UV and ozone equipment may be more sensitive to power quality events than other equipment (e.g., pumps) at a WTP and should be considered when evaluating historical events.

Backup and Power Conditioning Equipment

Power quality events can result in the AOT equipment shutting off, which may result in undertreated water. If infrequent but sustained power quality events are expected, a simple backup power supply (e.g., generator) may be sufficient. Existing backup power supplies should be evaluated to determine if they can accept the additional load because of the high load required for AOT systems.

Power conditioning equipment may be necessary if the power quality events are expected to be frequent and could result in undertreated water, especially for systems targeting disinfection as well as oxidation. Two examples of power conditioning equipment include (USEPA 2006):

- **Uninterruptible power supply (UPS)** systems provide continuous power in the event of a voltage sag or power interruption. The battery capacity supplies power to all connected equipment until a generator starts. There are both online and offline UPS systems that could be considered.

- **Active series compensators** boost voltage to the equipment by injecting a voltage in series with the remaining voltage during a sag condition. They protect electrical equipment against momentary voltage sags and interruptions; however, they cannot correct sustained power quality problems.

SITE CONSTRAINTS/LAYOUT

AOT systems vary in size and layout depending on technology and system operation. Ozone AOT systems generally require a larger footprint than UV AOT systems because of the longer required contact time; however, there are several key aspects common to AOT systems to consider when developing the AOT system layout and piping configuration.

- Configuration of the connection piping and the inlet/outlet piping necessary before and after the ozone contactor or UV reactors.
- Space and piping for booster pumps (if necessary).
- Space for electrical equipment, including control panels, transformers, and power conditioning equipment (if needed).
- Adequate distance between adjacent ozone generators or UV reactors to afford access for maintenance tasks.
- Room for storing spare parts and chemicals (if needed).
- Lifting capability for heavy equipment.
- Chemical storage facilities (i.e., tank, concrete pad, possible shade covering) and accessibility for chemical delivery.
- Hydrogen peroxide quenching equipment (if required).
- Safety considerations for spacing of equipment and/or chemicals.

UV AOT System Site Considerations

The following items should be considered when developing the UV reactor and piping configuration and facility layout:

- Number and configuration of the UV reactors.
  - Inlet/outlet piping
  - Reactors in series
  - Redundancy
- Orientation of the UV reactor (vertical or horizontal).
- Maximum allowable separation distance between the UV reactors and electrical controls.
- Chemical mixing upstream of UV reactors.
- Validation and inlet hydraulics as recommended by the UVDGM if disinfection credit is desired.

The UV manufacturer should be contacted to obtain reactor and control panel dimensions to help determine the facility footprint. The footprint and layout can vary depending on the number and size of reactors. Therefore, footprints and layouts should be estimated for all of the UV reactors being evaluated. The footprint and head loss through the facility can then be used to evaluate the feasible locations.

When designing an unfiltered surface water UV AOT facility layout, the site layout considerations are generally the same. However, one significant difference is the increased opportunity for debris, which can damage reactor components. UV AOT facility designs should incorporate features that prevent potentially damaging objects from entering the UV reactor (e.g., screens, baffles, or low-velocity collection areas) if debris is anticipated based on historical experience (USEPA 2006).

When designing an UV AOT facility for groundwater supplies, the most significant differences are access to the site and the potential for sand and gravel to enter the UV reactor. The AOT facility may need to be enclosed to limit unauthorized access. Bypass of the initial discharge from intermittently used wells can be considered to avoid scratching the quartz sleeves with the accumulated sand particles. A sand/debris trap or other removal equipment (e.g., basket strainers or cartridge filters) can be installed prior to the UV equipment if it is expected that sand can pass through the initial screen (USEPA 2006).

Ozone AOT System Site Considerations

Ozone AOTs using a concrete contactor typically have a larger footprint than UV systems and require space for ozone generators, ozone destruct units, cooling water systems, nitrogen boost systems, electrical gear, and chemical storage tanks (e.g., LOX storage and vaporization). Some ozone AOT systems use serpentine pipe contactors to reduce the facility footprint. Chemical storage tanks require space and accessibility for chemical delivery. It is also important to note that ozone poses a safety risk if present at high concentrations in the ambient air; therefore, there should be adequate ventilation in the facility.

The following items should be considered when developing the ozone contactor and facility layout:

- Contactor design (tank or pipeline)
- Contact time required for treatment
- Redundancy requirements for contactors and equipment
- Location of ancillary facilities
  - Ozone generators
  - LOX storage and vaporizers, if required
  - Nitrogen boost
  - Ozone generator cooling systems
  - Ozone destruct units

COST ESTIMATION

Identifying the most appropriate technology and location for an AOT system should include a cost evaluation. Preliminary life-cycle cost estimates typically include capital costs and annual O&M costs. Capital costs include the cost of the AOT equipment and piping, building (if necessary), pumping (if necessary), electrical and instrumentation provisions, site work, yard piping, contractor overhead and profit, pilot-testing (if necessary), engineering, legal, and permitting costs. Capital costs should also include hydrogen peroxide quenching and/or filter modifications (i.e., conversion to biofiltration), if required.

Annual O&M costs account for costs that are incurred on an annual basis and typically include estimated labor, energy, equipment replacement costs, and chemical usage (e.g., hydrogen peroxide or chlorine). The cost of labor, power, equipment replacement, and chemical usage depends on the technology selected, and this cost difference should be accounted for in the annual O&M costs. The life-cycle cost analysis may also consider nonmonetary preferences that the water utility may have, including

- Manufacturer experience and service record
- Labor requirements
- Environmental/sustainability impacts (e.g., carbon footprint)
- By-product formation potential
- Footprint and site constraints
- Flexibility for future treatment expansion
- Reliability

TREATABILITY TESTING

AOT systems have limited design tools available compared to disinfection applications. UV disinfection reactors have validation reports that provide confidence in the ability of UV reactors to deliver the necessary UV dose as a function of flow and water quality. UV AOT reactions are more complicated than disinfection applications, and validation information is not available. Similarly, ozone has well defined CT requirements for disinfection, but CT values may not available for all AOT target contaminants. Literature resources can be used to estimate ozone or UV doses for target contaminants. However, AOT performance is very site specific because of water quality variations and should be evaluated on a case-by-case basis. Equipment manufacturers may require water samples in order to properly design an AOT system to account for site-specific water quality (e.g., hydroxyl radical scavenging demand).

AOT systems are a proven effective treatment method; however, the overall process is dependent on numerous water quality and operation factors. Treatability testing can be a beneficial option to provide site-specific data concerning feasibility, technology selection, and operational optimization. Treatability testing can be done at the bench scale and/or pilot scale. Bench-scale testing typically entails a batch or semibatch reactor using the site's source water and can be performed at a manufacturer's facility or outsourced to a laboratory. Pilot-scale testing involves flow-through systems and is usually conducted at or near the AOT facility site.

For both UV and ozone AOT systems, bench-scale testing can provide valuable information regarding contaminant removal efficiency, by-product formation, and hydrogen peroxide quenching (chemical quenching only). Bench-scale testing is less expensive and logistically demanding than pilot-scale testing. Bench-scale testing results can be difficult to directly scale up with UV AOTs using the E_{EO} approach. However, some manufacturers are beginning to size full-scale systems by combining bench-scale data with computational fluid dynamics and intensity (CFD-I) modeling (Bircher et al. 2012). Bench-scale testing for ozone AOT can be a useful method for evaluating ozone and peroxide doses and evaluating by-product formation and potential mitigation strategies.

Pilot-scale testing is more expensive and logistically rigorous than bench-scale testing but holds several important advantages. Pilot-scale testing provides more thorough data regarding contaminant removal efficiency, by-product formation, and hydrogen peroxide quenching (both chemical and catalytic with GAC), as more data can be collected over a range of water qualities and operating conditions. Pilot-scale testing can also provide valuable operator experience with AOTs. Bromate mitigation strategies for ozone AOTs can also be more effectively evaluated at the pilot-scale depending on the selected ozone and peroxide dosing strategy. In the case of UV AOT systems, pilot-scale testing can provide information regarding lamp sleeve fouling.

The specific goals and protocol of treatability testing are largely project-specific, but the tests can be designed to address the following subjects and generate site-specific data to

- Evaluate AOT feasibility and operation
- Evaluate required doses for UV, ozone, and oxidants (e.g., chlorine or peroxide). Note that UV dose calculations can be difficult at the pilot scale, and utilities may have to rely on vendor models for dose calculations. UV AOT systems designed based on an EEO cannot scale up bench-scale or pilot-scale data due to differences in UV reactor configurations.
- Compare ozone AOT and UV AOT systems with regard to operability, cost, and performance
- Demonstrate achievability of target contaminant removals
- Evaluate regulated and unregulated by-product formation of AOT systems
- Evaluate bromate formation with ozone AOTs and determine level of mitigation that can be achieved
- Evaluate UV sleeve fouling with UV AOT systems

- Identify peroxide quenching methods, contact times, quenching effectiveness, and by-product formation
- Provide water system operators experience with AOT equipment and O&M requirements

REFERENCES

Bircher, K., M. Vuong, B. Crawford, M. Heath, and J. Bandy. 2012. Using UV Dose Response for Scale Up of UV/AOP Reactors. In *Proceedings IWA World Water Congress, Busan, South Korea*. London: International Water Association.

Bolton, J.R. and C.C. Cotton. 2008. *The Ultraviolet Disinfection Handbook*. Denver, CO: American Water Works Association.

Collins, J., C. Cotton, and M. MacPhee. 2011. Advanced Treatment for Impaired Water Supplies: When Advanced Oxidation Systems Are the Best Option. *Water Practice & Technology*, 6(4).

Collins, J., J. Biggs, G. Maseeh, J. Dettmer, and C. Cotton. 2014. Staying Ahead of the Curve: Tucson Water's Management and Treatment of 1,4-Dioxane. In *Proc. Water Quality Technology Conference, New Orleans, LA*. Denver, CO: American Water Works Association.

Cotton, C.A., L. Passantino, D.M. Owen, M. Bishop, M. Valade, W. Becker, R. Joshi, J. Young, M. LeChevallier, and R. Hubel. 2005. *Integrating UV Disinfection Into Existing Water Treatment Plants*. Denver, CO: Awwa Research Foundation and American Water Works Association.

Dotson, A.D., V.S. Keen, D. Metz, and K.G. Linden. 2010. UV/H_2O_2 Treatment of Drinking Water Increases Post-Chlorination DBP Formation. *Wat. Res.*, 44:3707–3713.

Festger, A. 2014. UV-Oxidation for Recalcitrant Chemical Contaminants: Successes, Challenges and Future Applications. In *Proc. IUVA America's Regional Conference, White Plains, NY*. Florence, KY: International UV Association.

IEEE 519—IEEE Recommended Practice and Requirements for Harmonic Control in Electric Power Systems. Piscataway, NJ: IEEE.

Keen, O.S., A.D. Dotson, and K.G. Linden. 2013. Evaluation of Hydrogen Peroxide Chemical Quenching Agents following an Advanced Oxidation Process. *J. Environ. Eng.*, 139:137–140.

Lauderdale, C., J. Brown, P. Chadik, and M. Kirisits. 2011. *Engineered Biofiltration for Enhanced Hydraulic and Water Treatment Performance*. Denver, CO: Water Research Foundation.

NWRI. 2003. *Ultraviolet Disinfection Guidelines for Drinking Water and Water Reuse*, 3rd Ed. Fountain Valley, CA: National Water Research Institute in collaboration with American Water Works Association Research Foundation.

Randtke, S., and M. Horsley. 2012. *Water Treatment Plant Design*, 5th Ed. New York: AWWA, American Society of Civil Engineers, and McGraw-Hill.

USEPA. 2006. *Ultraviolet Disinfection Guidance Manual for Final Long Term 2 Enhanced Surface Water Treatment Rule*. Washington, DC: Office of Water. http://www.epa.gov/safewater/disinfection/lt2/pdfs/guide_lt2_uvguidance.pdf.

US Peroxide. 2015, March 17. Hydrogen Peroxide (H_2O_2) Safety and Handling Guidelines [Online]. http://www.h2o2.com/technical-library/default.aspx?pid=66&name=Safety-amp-Handling. Accessed March 17, 2015.

Venkatesan, N., G. Hua, D.A. Recknow, and K. Kjartanson. 2003. Impact of UV Disinfection on DBP Formation From Subsequent Chlorination. In *Proc. Water Quality Technology Conference, Philadelphia, PA*. Denver, CO: American Water Works Association.

9

Start-up, Operations, and Maintenance[1]

There are numerous components comprising an AOT system with each requiring tailored start-up procedures as well as operations and maintenance tasks. These components include the AOT system (i.e., UV or ozone), chemical feed systems, and online monitoring equipment. This chapter describes the start-up, annual O&M, monitoring, recording, and reporting procedures that can be used to confirm that the AOT system is operating properly.

The USEPA has provided suggested recommendations for the start-up, operations, and maintenance for some components of AOT systems (i.e., UV and ozone processes). The USEPA recommendations are for UV and ozone in disinfection applications, but many of the concepts can be adapted to apply to UV or ozone in an AOT context. Furthermore, for chemical feed systems and monitoring instrumentation, the start-up, operations, and maintenance are generally specific to the manufacturer. The governing agency should also be contacted to determine any specific requirements for annual O&M tasks, reporting needs, and required permitting and submittals.

STEPS FOR FACILITY START-UP

During the construction phase, prior to facility start-up, all system components must be installed properly according to the manufacturer's specifications. Installation should be coordinated with the manufacturers and verified by the contractor and engineer.

Once the AOT system and necessary facilities are constructed, several steps should be completed to confirm that the facility is ready for full-scale operation and any additional governing agency requirements. Start-up activities typically include the following:

- Coordination with the appropriate governing regulatory agency
- Development of a facility O&M manual
- Flushing and disinfecting the piping, pumps, valves, and equipment
- Hydrostatic testing

---

1 Some of this chapter has been adapted (with permission) from Bolton and Cotton (2008).

- Input/output and control loop testing
- Functional, performance, and commissioning testing
- Staff training

Once the AOT system is designed and the governing agency specific requirements are identified, an O&M manual can be developed. The procedures in the O&M manual should minimize the time the system is off-line and the possibility of system malfunction. The O&M manual should contain the following items (adapted from USEPA 2006):

- General description of the AOT system and components
- Relationship to other unit treatment processes
- Design criteria
- Control strategy and monitoring
- Monitoring, recording, and reporting
- Standard operating procedures
- Start-up procedures
- Shutdown procedures (manual and automatic)
- Safety issues
- Standard operating procedure for lamp breakage and mercury release (if applicable)
- Emergency procedures and contingency plan
- Alarm response plans
- Preventive maintenance needs and procedures
- Equipment calibration needs and procedures
- Troubleshooting guide
- Contact information for equipment manufacturers and technical services

Once construction is complete, functional and performance testing should be completed on all of the equipment associated with the system. Functional testing is intended to ensure everything is installed properly and all control systems and electrical work properly communicate with their intended counterparts. Functional testing can be considered a method to ensure that all systems are installed, connected, and linked together correctly prior to operation. Performance testing verifies that the AOT equipment and all of the components of the AOT system operate in conjunction with one another. The manufacturers of the various system components should be involved in the testing to confirm the systems are installed as specified.

The governing agency should be contacted to determine if water produced during functional and performance testing can be sent to the distribution system or if the water must be sent to waste. The duration of the testing is site specific and should demonstrate to the water utility and the governing agency that the facility can meet all applicable requirements during continuous operation.

Functional testing and performance testing are often the times of highest risk of damage to the equipment and impacts to the finished water quality. For example, UV reactors have a specific flow rate limitation based on the allowable bending stress on the UV lamp sleeves. High flow events can occur during start-up testing due to lack of coordination with the contractors and plant staff, equipment that is not installed properly, or failure of I&C controls. Detailed planning of start-up activities and close coordination with all staff on-site can help to reduce risks to the equipment, water quality, and staff. For chemical feed systems, function testing is typically completed with water to minimize the risk of chemical leaks/spills if the equipment was not installed properly. Once the chemical feed system functionality has been tested, the neat chemical can be used.

UV AOT functional testing may include the following items (adapted from USEPA 2006):

- Operation of individual UV reactors in automatic and manual modes to verify that the control system is functioning as intended
- Operation of the chemical feed system in automatic and manual modes to verify that the control system is functioning as intended
- Demonstration of UV reactor start-up and shutdown sequence
- Demonstration of response to alarms
- Measurement of electrical service voltage, current, and power consumption with different flow and water quality combinations to optimize energy use
- Assessment of the effectiveness of the cleaning system by inspecting sleeve transmittance and condition at regular intervals throughout the test period
- Confirmation that the programmed cleaning frequency correlates with the actual frequency of cleaning
- Confirmation of duty UV sensor accuracy using reference UV sensors
- Verification of the calibration of the online UVT analyzer

Ozone AOT functional testing may include the following items:

- Operation of individual ozone generators in automatic and manual modes to ensure the control system is operating as intended
- Operation of the chemical feed system in automatic and manual modes to verify that the control system is functioning as intended
- Demonstration of ozone generator start-up and shutdown sequence
- Efficiency monitoring of ozone generation
- Demonstration of response to alarms

- Measurement of electrical service voltage, current, and power consumption with different flow and water quality combinations to optimize energy use
- Verification of the calibration of the ozone monitoring equipment, including the ambient air ozone monitoring system

In addition to the individual UV or ozone system testing, the overall AOT system operation needs to be functionally verified to ensure that all components are working in unison. Functional testing of the entire AOT system can include the following items:

- Operation of the complete AOT system, including chemical feed systems, in automatic mode to ensure the control and dosing systems are working as intended
- Demonstration and observation of the system, including modulation of power and chemical dosing in response to flow and water quality changes
- Demonstration of standby train operation due to flow or water quality changes and alarms
- Start-up and shutdown of the complete AOT system to ensure proper procedures
- Verification of the system monitoring equipment
- Confirmation of backup generator and/or UPS power transfer to the AOT equipment (if applicable)

Before the testing is completed, the inventory of spare parts should be checked against the requirements described in the O&M manual. By maintaining the proper inventory of spare parts, maintenance activities of the AOT system can be performed in a timely and effective manner.

Staff training should also be conducted prior to the completion of testing. Proper staff training and staffing levels can minimize AOT system faults by helping the staff to properly maintain the equipment and be able to react in a timely manner. Proper staff training is also required for the safety of all staff members. AOT systems can present health and safety risks due to electrical, UV light, and chemical exposure. Chapter 11 provides additional details on the safety concerns with AOT systems. The staff level requirements are site and equipment specific.

Once the system is verified to be functional and operating as intended, performance testing should take place to ensure the system can, if required, be operated to achieve the desired contaminant removal levels and peroxide quenching. It is likely that the design conditions for flow, target contaminant concentrations, and water quality will not all be available during performance testing. The performance test planning should carefully consider methods to evaluate the hydraulic capacity and treatment goals given site-specific limitations and water disposal options. To verify operation at the design conditions, contaminant spiking and/or temporary water quality adjustments (e.g., pH or UVT adjustment) may be required. Testing water may not be allowed to be discharged as finished water because the AOT system is not yet verified, especially if any water quality adjustments are required. A strategy should be developed to either collect water for disposal or recycle it back through the plant so that water is not discharged.

The AOT system vendor is responsible for proving the operation and reliability of the system. Coordination with the vendor and contractor is important during performance testing to ensure the system operation can be properly verified as part of the overall treatment strategy and water system operation. The governing agency should also be included in this phase of the testing.

It is important to note that many system components have specific installation, calibration, and testing procedures provided by the manufacturer or in the construction specifications. These procedures, if necessary, should be performed with coordination by the manufacturer. Examples of these components include pumps, tanks, and monitoring sensors or fixtures. There are often procedures included for calibration and testing of these components, and the manufacturer should be contacted if no such information is included.

Once the functional testing and performance testing are completed, some systems may include a commissioning period as required by the equipment procurement or regulatory requirements. The governing agency may require a commissioning period (e.g., 30 or 90 days) to prove that the system can operate reliably before providing an operating permit.

ROUTINE OPERATIONS AND MAINTENANCE TASKS

Unlike disinfection applications, many AOT applications may not have specific regulatory recommendations/requirements for monitoring or maintaining AOT equipment; however, proper maintenance can reduce the potential for system malfunction or inadequate performance that can potentially lead to compliance issues. The manufacturer-specific annual O&M tasks should be followed to maintain operation of the system. Many of the maintenance activities are specific to the technology's manufacturer, and the information should be obtained from the equipment vendors.

UV AOT Systems

The USEPA UVDGM (USEPA 2006) provides recommended maintenance tasks for UV disinfection systems, and many of the tasks and frequencies can also be applied to UV AOT systems to ensure proper operation and maintenance of the UV system. Examples of UV system maintenance tasks and recommended frequencies are found in Table 9-1. However, not all tasks are required or recommended for all AOT systems and should be evaluated for applicability to a specific system and discussed with the regulatory agency. As such, the UV manufacturer should be consulted when determining the maintenance schedule. If the system is used for AOT operation and disinfection credit, the USEPA UVDGM (USEPA 2006) should be referenced.

Table 9-1 Example operations and maintenance tasks* (adapted from USEPA 2006)

Equipment	Frequency	Task
UV reactors	Daily	Perform overall visual inspection of the UV reactors.
		Confirm that system control is on automatic mode (if applicable).
		Check control panel display for status of system components and alarm status and history.
		Review 24-hour monitoring data to confirm that the reactor has been operating within design values during that period.
	Monthly	Check reactor housing, sleeves, and wiper seals for leaks and replace any damaged components.
UV lamps	Monthly	Check lamp run time values. Consider changing lamps if operating hours exceed design life.
	Bimonthly	Check intensity of UV lamps. Sharp decrease in lamp intensity may indicate lamp sleeve fouling.
	Lamp/manufacturer specific	Replace lamps when any one of the following conditions occurs: • Initiation of low UV intensity or low dose alarm after verifying that this condition is caused by low lamp output. • Initiation of lamp failure alarm after verifying it is not a nuisance alarm.
	When lamps are replaced	Send spent lamps to a mercury recycling facility or back to the manufacturer.
Ballasts	Daily	Verify that ballast cooling fans are operational and that ballasts are not overheated.
	Semiannually	Check ballast cooling fans for unusual noise.
	Manufacturer's recommended frequency	Check the ballast cooling fans for dust buildup and damage. Replace if necessary. Replace air filters (if applicable).
UV lamp sleeves	Sleeve/manufacturer specific	Replace sleeve when damage, cracks, or irreversible fouling significantly decreases UV intensity of an otherwise acceptable lamp. Inspect sleeves per the cleaning system inspection frequency below for damage and cracks. Adjust the replacement frequency based on operational experience.
Cleaning system	Weekly	Initiate manual operation of wipers (if provided) to verify proper operation.
	Semiannually	Check cleaning efficiency by recording the UV sensor reading before and after cleaning if fouling is observed.

Table continues next page

Start-up, Operations, and Maintenance   117

Table 9-1 Example operations and maintenance tasks* (adapted from USEPA 2006) (continued)

Equipment	Frequency	Task
Cleaning system (cont.)	Semiannually (online chemical cleaning systems)	Check cleaning fluid reservoir (if provided) and replenish as needed. Drain and replace solution if the solution is discolored.
	Manufacturer's recommended frequency	Inspect and maintain cleaning system drive as recommended by the manufacturer.
Duty UV sensors (if required for UV reactor control)	When duty UV sensors fail calibration	Send the duty UV sensors to a qualified facility (e.g., manufacturer) for calibration, or replace the duty UV sensors.
Reference sensor	Annually	Calibrate reference UV sensor by sending it to a qualified facility that uses a traceable standard [e.g., National Institute of Standards and Technology (NIST)].
UVT analyzer	Manufacturer's recommended frequency	Clean and replace UVT analyzer parts according to manufacturer's recommended procedure.
	Weekly	Verify calibration of UVT analyzer against a bench-top UV spectrophotometer.
Thermometer and/or water level monitor	Manufacturer's recommended frequency	Visually inspect and replace at the manufacturer's recommended frequency.
Online analyzers and flowmeters	Daily	Verify that all online analyzers, flowmeters, and data recording equipment are operating normally.
Ground Fault interrupter (GFI)	Annually	Test trip and maintain GFI breakers in accordance with manufacturer's recommendations.
Valves	Semiannually	Check operation of automatic and manual valves.

* Maintenance activities should be consistent with manufacturer's instructions.

Ozone AOT Systems

USEPA recommended annual O&M tasks for ozone systems are less defined than for UV systems (USEPA 1999); however, system components still require routine inspection and maintenance to ensure proper function. Ozone system maintenance is dependent on the installed equipment and staff training by the vendor should include all of the recommended maintenance tasks and frequencies. Typical inspection and maintenance tasks include

- Continuous monitoring of the ozone generator when in operation
  - Clean feed gas should be supplied to the generator at all times with a dew point of −60°C or lower, as moisture can initiate corrosion in the ozone generator

- Ensure generator cooling water flow is maintained and the generator temperature is monitored
- After ozone generator shutdown, maintain the unit in a pressurized state with a small flow of dry air or oxygen to eliminate moisture accumulation in the assembly
- Pump, compressor, or blower lubrication and service according to manufacturer's recommendations
- Replace filters periodically, if applicable, depending on air quality and system usage
- Periodic pressure testing of LOX tanks, if applicable
- Regularly inspect piping and contact chambers for leaks or corrosion
- Periodic dielectric tube cleaning, especially if generator efficiency drops
- Verify calibration of online analyzers (typically annually for ozone gas monitors and weekly for ozone residual analyzers)
- Maintenance and replacement of ozone diffusers (typically annually), if applicable
- Periodic descaling of heat exchanger (based on pressure drop across heat exchanger) if operating a closed-loop system with heat exchanger and depending on water quality
- Periodic testing of the closed-loop cooling water quality if operating closed-loop cooling of ozone generators

Ozone systems should be actively monitored for decreases in efficiencies as that may indicate the system requires maintenance. However, because moisture can be detrimental to the system, the generators and dielectrics should not be opened and cleaned without significant reason. The ozone system manufacturer should include detailed operations and maintenance procedures and frequencies and provide support for some maintenance tasks.

Oxidant Injection System

Annual O&M tasks for an oxidant injection system (e.g., hydrogen peroxide or chlorine) are specific to the manufacturer and dependent on system operation. Injection systems typically consist of chemical storage tanks, injection pumps, associated piping, and instrumentation.

General storage tank preventative maintenance tasks include:
- Inspection of the tank for physical damage including leaks, cracks, cuts, bulges, or softening
- Inspection of the tank fittings, gaskets, and valves for any wear, deterioration, damage, or leaks

Storage tanks are generally fitted with various sensors such as pressure transmitters, level sensors, and temperature sensors to monitor operation. These sensors should be installed, calibrated, and maintained according to the manufacturer's guidelines.

General preventative maintenance tasks for chemical metering pumps include the following:

- Inspection of pumps for physical damage such as leaks, cracks, or deterioration
- Proper pump lubrication
- Diaphragm or tubing replacement per manufacturer's recommendations

The pump should be properly calibrated to ensure an accurate flow rate. Some systems may use a flowmeter for the pump, which also requires proper installation, calibration, and maintenance according to the manufacturer's guidelines. Other systems rely on drawdown tests to verify the metering pump flow rate.

The system should be carefully observed throughout start-up and operation to determine the frequency for specific maintenance tasks and to develop a detailed maintenance plan for each component. Utility personnel should become familiar with the manufacturer's operation, calibration, and maintenance procedures to troubleshoot issues and ensure effective system operation. Staff training should include training on all annual O&M tasks of the chemical injection systems.

Hydrogen peroxide and sodium hypochlorite are prone to off-gassing, which can result in gas and bubbles entering the chemical metering pump. Gas buildup can bind the pumps or piping/tubing and prevent chemical injection. A strategy should be developed to bleed gas from the system and should be optimized during initial operation. Off-gas control strategies may include selection of pumps with automatic off-gassing or initially operating the pumps at a higher than required flow rate to bleed off any bubbles and strategic location of degassing valves within the chemical feed piping. If increased pumping speeds are used, the downstream implications for a short-term higher oxidant residual should be carefully considered (e.g., increased quenching requirement).

MONITORING

AOT systems require that several system parameters be monitored to ensure proper system functioning. This includes monitoring of the UV or ozone system, chemical injection equipment, and quenching process (if applicable).

As opposed to UV disinfection, which uses online monitoring of certain parameters to verify adequate UV dose delivery for regulatory compliance, AOT systems typically depend on the target contaminant concentration in the finished water to monitor regulatory compliance. Therefore, the monitoring regime for AOT systems is different from that for disinfection systems. If the AOT system is used both for disinfection and contaminant oxidation, monitoring for disinfection credit will likely be the driver for monitoring tasks and frequencies.

UV AOT Systems

An adequate E_{EO} or UV dose should be applied to create the target hydroxyl radical concentration. The E_{EO} or UV dose is a function of power input or UV irradiance, flow rate, and UVT. However, the E_{EO} or UV dose is not used for regulatory purposes in UV AOT systems as it is in UV disinfection. Therefore, monitoring may be less stringent for parameters vital to E_{EO} or UV dose determination and regulatory compliance. However, it is still important to monitor the parameters as a measure of system performance and to determine the need for any maintenance.

UV Sensor Calibration

Although hydroxyl radical formation is dependent on the UV lamp irradiance, the specific irradiance of each lamp is less important for AOT systems than for UV disinfection as compliance is based on the treated water contaminant concentration. Therefore, the UV AOT system is not as dependent on UV irradiance monitoring for regulatory compliance. However, the UV lamp irradiance is an important parameter as it can monitor UV system operation and efficiency and provide insight into issues such as fouling and aging. The UV irradiance is typically monitored per lamp, bank of lamps, or per reactor depending on the system configuration. The calibration of UV sensors should be verified as with other online instrumentation. For disinfection applications, the USEPA recommends the calibration of the UV sensors be evaluated by comparing the reading of the duty UV sensors to that of a reference UV sensor using Eq. 9-1, and these same methods can be used for AOT systems.

$$\left(\frac{S_{Duty}}{S_{Ref}}\right) \leq 1.2 \qquad (Eq.\ 9\text{-}1)$$

Where:

S_{Duty} = intensity measured with the duty UV sensor (mW/cm²)

S_{Ref} = intensity measured with the reference UV sensor (mW/cm²)

The importance of UV sensor calibration is also vendor specific depending on the control strategy. If operation is based on a target E_{EO}, the individual UV sensor readings may not be used to determine the operating conditions, as operation of the equipment is contingent on electrical energy input. However, UV sensor readings may be directly used if the system operation is based on a target UV dose. If the UV sensor readings are used for reactor control, the calibration of the UV sensors should be periodically verified based on manufacturer and governing agency requirements. As a point of reference, the calibration of duty UV sensors must be verified on a monthly basis for disinfection applications.

UVT Analyzer Calibration

UVT is an important control parameter for UV AOT systems and is directly used to determine operating conditions for systems. Therefore, UVT analyzers must remain calibrated. For disinfection applications, the USEPA recommends the calibration of online UVT analyzers be evaluated by comparing the reading of the online UVT analyzer to that of a calibrated bench-top spectrophotometer using Eq. 9-2. If the UVT analyzer is found to be out of calibration, the analyzer should be recalibrated.

$$\left| \text{UVT}_{\text{on-line}}(\%) - \text{UVT}_{\text{bench}}(\%) \right| \leq 2 \text{ percent UVT} \quad \text{(Eq. 9-2)}$$

Where:
$\text{UVT}_{\text{online}}$ = UVT measured by the online UVT analyzer (%)
$\text{UVT}_{\text{bench}}$ = UVT measured by a bench-top spectrophotometer (%)

The USEPA recommends that online UVT analyzers be monitored at least weekly for disinfection applications due to the importance of UVT in dose calculations and the inherent variability of online UVT analyzers. A similar monitoring frequency should be considered for AOT applications during initial operation. The monitoring frequency can be adjusted based on the calibration results obtained over the first year of operation and as approved by the governing agency.

UVT analyzers can be calibrated on-site using the UV manufacturer's recommended approach if it fails the criterion in Eq. 9-2. If the UVT analyzer cannot maintain calibration for 24 hours, most UV AOT systems will allow for manual UVT inputs. Selection of the manual UVT input can be based on current operation or a conservative value.

Monitoring and Recording Frequencies

As previously stated, direct monitoring of AOT system performance is not well defined and is dependent on the requirements of the governing agency. However, typical monitoring strategies can be adapted from the USEPA UVDGM as guidelines to ensure proper function and calibration of the UV equipment. Example monitoring and recording frequencies for monitoring parameters are shown in Table 9-2.

Ozone AOT Systems

As with UV AOT systems, direct monitoring of ozone AOT system performance is not well defined and depends on the requirements of the governing agency. However, typical monitoring strategies are similar to those for UV AOT systems. Example monitoring and recording frequencies for monitoring parameters are shown in Table 9-3.

Oxidant Injection

The oxidant injection and quenching system, if required, will also require effective monitoring to verify treatment performance. Deviations in oxidant dosing can result

Table 9-2 Example monitoring frequencies for key UV AOT operational parameters (adapted from USEPA 2006)

Parameter	Example Monitoring Frequency
UV intensity	Continuous
UVT	Continuous
Lamp status	Continuous
Flow rate	Continuous
Production volume	Continuous
Calibration of UV sensors	Monthly (depending on inclusion of UV intensity in control strategy)
Calibration of online UVT analyzer	Weekly*
Power draw	Continuous
Water temperature	Continuous
UV lamp on/off cycles	Continuous
Hydroxyl radical scavenging demand parameters (pH, TOC, alkalinity)	Depends on source water variability
Target contaminant influent and effluent concentrations	Depends on source water variability and regulatory requirements
Operational age† of the following equipment: • Lamp • Ballast • Sleeve • UV sensor	Monthly

* Frequency could be reduced depending on monitoring results.
† The operational age is the amount of time the equipment was running that month.

in variations in contaminant destruction. Online instrumentation can be considered for monitoring oxidant injection and residuals. The accuracy and reliability of online analyzers depend on the selected oxidant. A range of online free chlorine analyzers available with proven accuracy and reliability. However, online hydrogen peroxide analyzers are more limited and require routine monitoring and maintenance to maintain calibration. Monitoring of online analyzer calibration is critical for any systems to actively use the online instrumentation for system control. The analyzers will also require periodic maintenance for adding reagents or maintenance of analyzer probes. Maintenance and calibration of the online instrumentation should be performed as per manufacturer recommendations.

ENERGY AND CHEMICAL MANAGEMENT

Operating costs for AOT systems are generally higher than disinfection applications due to the additional energy and chemical requirements. System designers and operators

Table 9-3 Example monitoring frequencies for key ozone AOT operational parameters

Parameter	Example Monitoring Frequency
Ozone feed gas concentration	Continuous
Applied ozone dose	Continuous
Ozone residual	Continuous
Ozone generator cooling water temperature	Continuous
Flow rate	Continuous
LOX usage	Continuous (if applicable)
Feed gas moisture content	Continuous
Calibration of ozone residual monitors	Weekly*
Power draw	Continuous
Water temperature	Continuous
pH	Continuous
Hydroxyl radical scavenging demand parameters (pH, TOC, alkalinity)	Depends on source water variability
Target contaminant influent and effluent concentrations	Depends on source water variability and regulatory requirements
Raw water bromide and finished water bromate	Monthly

* Frequency could be reduced depending on monitoring results.

should tailor the design based on source water quality characteristics and target contaminants and concentrations. This can be accomplished through adequate source water quality monitoring, bench-scale or pilot-scale testing, and system performance observation during the initial start-up and operation of the system. Testing and observation should consider the following:

- Hydroxyl radical scavenging demand
- Water quality characteristics (pH, TOC, bromide, nitrate, etc.)
- Hydrogen peroxide use
- Energy use of the system

For most AOT systems, the vendors will guarantee maximum energy and chemical dose(s) required to meet the design criteria, and these parameters need to be verified during performance testing. The performance testing establishes an operational baseline. During operation, the cost of power versus the cost of chemicals can be evaluated to determine if operation of the system can be optimized from the conditions established during performance testing. Systems with high power costs may benefit from lower energy use and higher chemical doses. For many installations, power will be cheaper than chemicals, and thus the system will often run at a high power to limit chemical usage.

Limiting chemical usage can also be beneficial if the residual after treatment must be quenched.

REFERENCES

Bolton, J.R. and C.C. Cotton. 2008. *The Ultraviolet Disinfection Handbook*. Denver, CO: American Water Works Association.

USEPA. 1999. Chapter 3 Ozone. *EPA Guidance Manual Alternative Disinfectants and Oxidants*. Washington DC: Office of Water. US Environmental Protection Agency.

USEPA. 2006. *UV Disinfection Guidance Manual*. Washington DC: Office of Water. US Environmental Protection Agency.

10

AOT Case Studies

This chapter contains example case studies as illustrations of practical AOT applications.

TREATMENT OF TASTE AND ODOR[1]

Blue-green "algae"[2] blooms occur in bodies of water where there are sufficient nutrients in the water during the summer. When the algae die, the decaying organic matter produces compounds that give an earthy/musty, fishy, swampy, grassy taste and odor to the drinking water; the principal compounds [plus other taste-and-odor (T&O) compounds] are geosmin and 2-methylisoborneol (2-MIB). These compounds are not toxic, but are unpleasant and unappealing to the public. However, under certain conditions, very toxic compounds (e.g., microcystin and anatoxin) can be produced.

The by-products of blue-green algae blooms have been treated by various treatment technologies, including

- Potassium permanganate—weak oxidant, limited effectiveness.
- Powdered activated carbon (PAC)—messy, increased sludge handling.
- Granular activated carbon (GAC)—frequent and expensive change-outs.
- Ozone or ozone/H_2O_2—effective for T&O with simultaneous disinfection; however, it can be a more complex system than UV-advanced oxidation with concerns for by-product formation (bromate). Addition of hydrogen peroxide can increase treatment capacity.
- UV/H_2O_2 or UV/Cl_2 – effective for T&O with simultaneous disinfection.

K-Water of South Korea is an example utility that has recently installed AOTs for control of T&O compounds. K-Water owns and operates the Sung-Nam Water Treatment Plant (WTP) and Si-Heung WTP. Both WTPs use the same water source (i.e., Pal Dang Dam Reservoir) that experiences seasonal T&O events with geosmin and 2-MIB. South Korea decided to invest more than $70 billion to build additional advanced water treatment facilities using AOTs in combination with activated carbon filtration. This

---

1 This section is based on material provided by WEDECO, a Xylem Brand of Herford, Germany.
2 Blue-green "algae" are really photosynthetic bacteria.

was a driving factor for K-Water to implement an AOT system at both WTPs to effectively remove T&O compounds and other contaminants of emerging concern to ensure the delivery of high quality drinking water. Both WTPs were conventional plants with coagulation, sedimentation, and dual-media sand filtration. UV aned ozone AOT were considered for both plants and ultimately ozone/peroxide was selected for the Sung-Nam WTP and UV/peroxide was selected for the Si-Heung WTP. The reasons for selection include

- Sung-Nam WTP
  - Space had already been reserved for advanced treatment process allowing ozone to be installed prefiltration.
  - Capital cost for ozone/H_2O_2 was lower compared to UV/H_2O_2 due to the higher design flow rate (218 mgd for the Sung-Nam WTP versus 28 mgd for the Si-Heung WTP).
- Si-Heung WTP
  - Preozonation would require purchasing additional property due to the footprint of ozone treatment. Securing more property would have taken at least two years and would likely delay construction. UV/H_2O_2 did not require additional property to be purchased.
  - Clean source water allowed the UV/H_2O_2 AOT to be cost-effectively installed upstream of filtration.

The design parameters for each WTP are provided in Tables 10-1 and 10-2. Figures 10-1 and 10-2 illustrate the full-scale installations.

As of 2015, the Sung-Nam WTP has not had a T&O event to verify performance of the ozone/H_2O_2 system. However, the UV/H_2O_2 system at the Si-Heung WTP has undergone performance testing to evaluate the overall performance of the system and to establish the control logic based on testing results. The performance testing showed that the UV/H_2O_2 system was capable of reducing 2-MIB concentrations by up to 87 percent depending on the lamp power setting and hydrogen peroxide dose (Figure 10-3).

TREATMENT OF MICROPOLLUTANTS

Currently, there are concerns about many micropollutants in drinking water, such as pesticides and herbicides, endocrine-disrupting compounds (EDCs), *N*-nitrosodimethylamine (NDMA), and 1,4-dioxane. Conventional water treatment technologies cannot treat many of these micropollutants, with the possible exception of ozone treatment. UV alone will not work (except for NDMA) because most of these compounds do not absorb UV. The following sections provide two examples of AOTs used for micropollutant treatment.

AOT Case Studies

Table 10-1 Sung-Nam Water Treatment Plant ozone/H$_2$O$_2$ design criteria

Design flow rate	218 mgd
Target log reduction	0.5-log MIB
Ozone dose	2 mg/L
Hydrogen peroxide dose	1 mg/L
Number of ozone generators	Three 2,700-lb/day generators feed by liquid oxygen
Ozone contactor	Two parallel concrete contact tanks

Figure 10-1 Sung-Nam Water Treatment Plant ozone generator and hydrogen peroxide storage system (courtesy of WEDECO, a Xylem Brand).

O$_3$/H$_2$O$_2$ AOT for Pesticide Treatment[3]

Agriculture is one of the major industries in the regions served by Anglian Water in the United Kingdom. Consequently, the raw water for potable supply is often challenged by pesticides. Since the early 1990s, conventional treatment for surface water plants has relied on the combination of ozone and GAC to achieve the European Union–legislated pesticide compliance levels of 0.1 μg/L. However, in recent years there has been an increase in pesticides that are more difficult to treat. Especially prevalent, although not exclusively in the Anglian Water region, is metaldehyde. WTPs using conventional ozone designs would likely have difficulty meeting the regulatory limits for the pesticides and the legislated bromate levels of 10 μg/L. The addition of hydrogen peroxide to the process has the potential to manage bromate formation while increasing treatment. A pilot-scale evaluation was completed by Anglian Water to evaluate the performance of ozone/H$_2$O$_2$ for pesticide treatment, while minimizing bromate formation.

The pilot system included a 176-gpm ozone/H$_2$O$_2$ contactor at a surface water treatment plant treating filtered water. The ozone/H$_2$O$_2$ system was shown to be capable of achieving up to a 0.7-log reduction of metaldehyde with a applied maximum ozone dose of 10 mg/L (Figure 10-4). The system was able to maintain bromate formation below 5 μg/L with an influent bromide concentration of 80 μg/L.

---

3   This section is based on material provided by WEDECO, a Xylem Brand of Herford, Germany.

Table 10-2 Sung-Nam Water Treatment Plant UV/H$_2$O$_2$ design criteria

Design flow rate	28 mgd
Target log reduction	0.5-log MIB
Ozone dose	2 mg/L
Hydrogen peroxide dose	0.5 mg/L
Number of UV reactors	Three UV reactors in parallel
Lamp type	Low-pressure high-output
Number of lamps per reactor	168
Power per lamp	0.295 kW
Total power per reactor	161 kW

Figure 10-2 Si-Heung Water Treatment Plant UV reactor and hydrogen peroxide storage system (courtesy of WEDECO, a Xylem Brand).

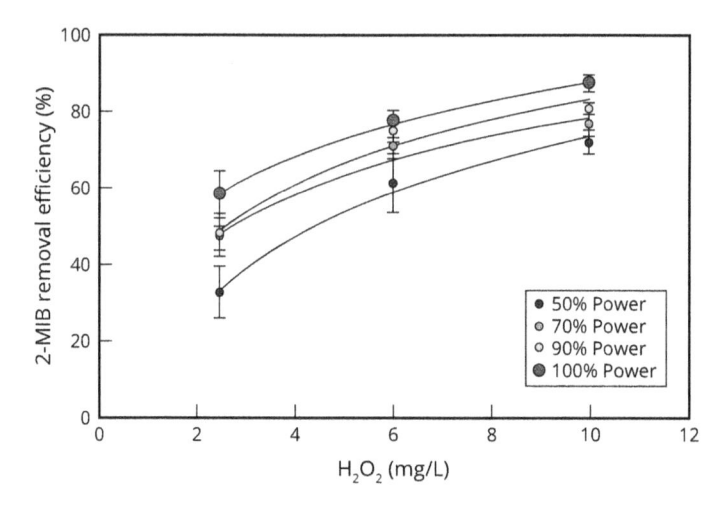

Figure 10-3 Si-Heung Water Treatment Plant UV/H$_2$O$_2$ full-scale performance test results (courtesy of WEDECO, a Xylem Brand).

AOT Case Studies 129

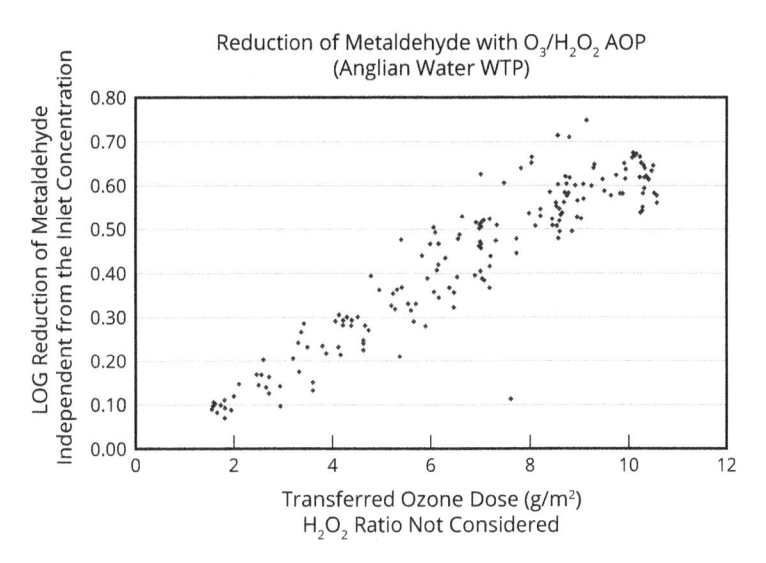

Figure 10-4 Metaldehyde reduction as a function of ozone dose (courtesy of WEDECO, a Xylem Brand).

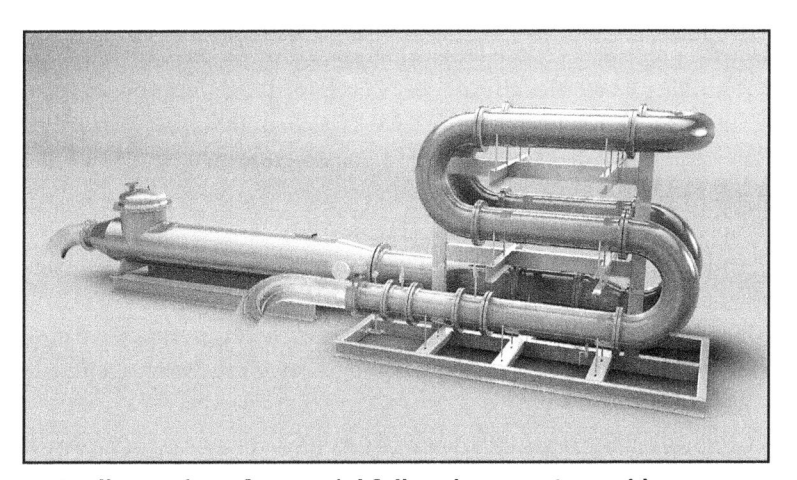

Figure 10-5 Illustration of potential full-scale ozone/peroxide system (courtesy of WEDECO, a Xylem Brand).

A peroxide dose of 16–22 mg/L was shown to maintain bromate formation below 3 μg/L. Figure 10-5 provides an example illustration of a potential full-scale ozone/H_2O_2 system that uses pipe contractors instead of a contactor tank design.

UV/H_2O_2 AOT for 1,4-Dioxane Treatment

The Tucson International Airport Area Groundwater Remediation Project (TARP) is an example of a UV AOT application for 1,4-dioxane treatment. The TARP wells and water treatment plant are owned and operated by Tucson Water (Tucson, Arizona).

Figure 10-6 UV reactors and GAC contactors used at the TARP AOP water treatment facility.

These facilities have been cleaning up TCE and other VOCs from groundwater at one of Arizona's largest federal superfund sites and providing Tucson Water's customers with high-quality drinking water since 1994.

In 2002, 1,4-dioxane was first detected in TARP groundwater because of advances in laboratory analysis methods. The existing treatment process at the TARP WTP, packed column aeration, is ineffective for 1,4-dioxane removal. The only proven municipal-scale water treatment process for 1,4 dioxane are AOTs.

The USEPA published a revised toxicological evaluation in 2010, and subsequently, reduced its Drinking Water Health Advisory Levels for 1,4-dioxane by nearly an order of magnitude in 2011. After publication of USEPA's revised advisory levels for 1,4-dioxane, Tucson Water commissioned the design and construction of a new $14.6M advanced oxidation process (AOP) water treatment facility to treat groundwater from the remediation wells upstream of the existing plant (Figure 10-6). The treatment process employs UV/H_2O_2 and GAC quenching for hydrogen peroxide removal. Table 10-3 provides the system design criteria and sizing.

The control system for the UV AOP equipment includes an automated optimization approach for all of the following variables based on basic criteria entered by operations staff:

- Partial flows can be bypassed around the AOP water treatment facility for blending with UV/H_2O_2 treated water upstream of the TARP WTP's packed columns. Bypass allows for optimized annual O&M costs by only treating the volume of water necessary to achieve the water quality goals. The bypass flow rate is optimized by the control system based on flow-weighted remediation well water quality data and the blended water 1,4-dioxane goal.

- The hydrogen peroxide dosage and UV lamp intensity are optimized by the control system based on hydrogen peroxide and electric power unit costs and the 1,4-dioxane treatment goal for the UV AOT reactors, thus minimizing unit treatment costs.

Transforming a significant regional groundwater quality problem into a high-quality drinking water supply has resulted in a range of positive social and economic impacts to the

Table 10-3 TARP UV/peroxide and GAC design criteria

Design flow rate	5,800 gpm
Target log reduction	1.6-log 1,4-dioxane
Design UVT	95%
Number of UV reactors	3 parallel trains with 2 UV chambers per train in series and 2 reactors per chamber
Lamp technology	Low-pressure high-output
Total connected load	222 kVA
Design hydrogen peroxide dose (maximum)	12 mg/L
GAC contactor type	Pressure contactors
Number of contactors	8 contactors in parallel
Contactor diameter	10 ft
GAC media	Catalytic coconut shell-derived carbon (12×40 mesh)
Minimum empty bed contact time	5 minutes
Target GAC treated water hydrogen peroxide concentration	<0.2 mg/L

Tucson community for the past 20 years TARP has operated. The original facilities stop further migration of VOC contamination in the regional aquifer and are restoring this important water resource in a semi-arid community where all available water resources are critical to economic development and the local quality of life. The new AOP water treatment facility enables full use of the original TARP remediation wells, restoring the facilities to full effectiveness without compromising water quality commitments or requiring additional water supplies for blending.

The design and reliable operational results for the full-scale processes were facilitated by pilot testing of both the UV/H_2O_2 and the GAC hydrogen peroxide quenching system. The pilot testing program allowed

- Verification and demonstration of achieving aggressive 1,4-dioxane destruction, to below USEPA Drinking Water Health Advisory Levels.

- Verification, media selection, and design criteria refinement for hydrogen peroxide quenching with GAC, for which there was little precedent.

The AOP water treatment facility project was completed in 2014 and has met or exceeded Tucson Water's needs and expectations in terms of 1,4-dioxane treatment performance, restoration of full operation of the remediation well fields and elimination of blending requirements with other supplies, operational efficiency, schedule, and cost. The AOP water treatment facility has also been successful in simultaneously reducing TCE concentrations to below the method reporting limit of 0.5 μg/L. Figures 10-7 through 10-9 present example treatment and H_2O_2 removal performance that has been representative of full-scale operation.

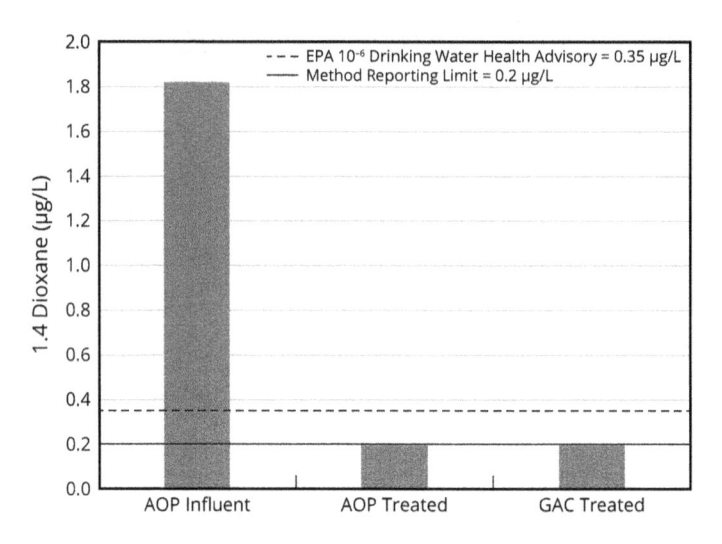

Figure 10-7 1,4-dioxane removal through UV/H$_2$O$_2$ treatment.

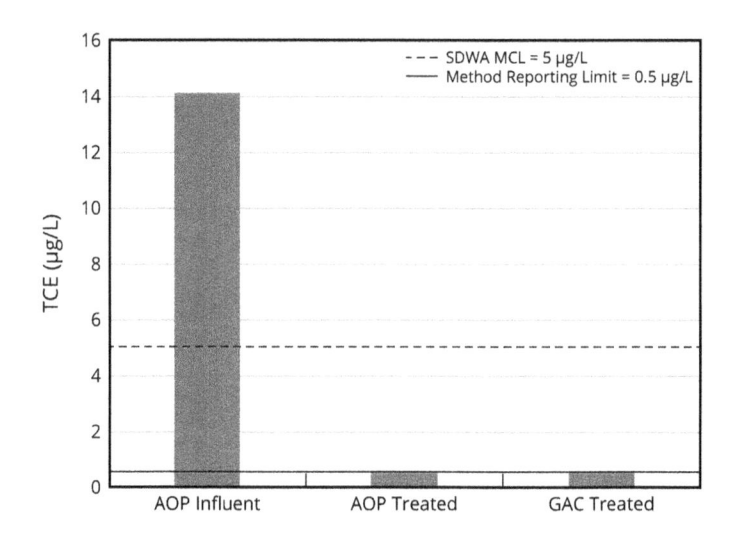

Figure 10-8 TCE removal through UV/H$_2$O$_2$ and GAC.

REUSE TREATMENT[4]

AOTs can have multiple applications within a wastewater or wastewater reuse applications. One of the most common applications is for reducing NDMA concentrations NDMA is an unregulated contaminant that is found in many wastewaters. Despite the

---

4 This section is based on material provided by Calgon Carbon of Pittsburgh, USA.

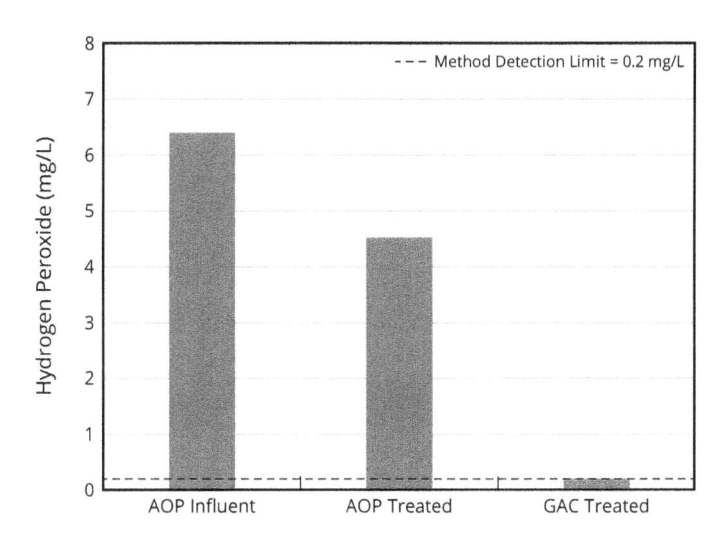

Figure 10-9 Hydrogen peroxide removal through UV/H$_2$O$_2$ and GAC.

Table 10-4 City of Scottsdale AWT UV photolysis design criteria

Design flow rate	20 mgd
Target log reduction	1.0-log NDMA
Design UVT	95%
Number of UV reactors	1 (48-in. reactor with 18 lamps)
Lamp technology	Medium pressure
Power per lamp	20 kW
Total connected load	396 kW
Design hydrogen peroxide dose (maximum)	0 mg/L

fact that there is no federal regulatory limit, many states have adopted standards for drinking water and reuse applications due to the potential health effects associated with NDMA. NDMA is often formed within wastewater treatment plants due to the reaction of chloramines with NDMA precursors.

The City of Scottsdale, Arizona, Water Campus is one of several example reuse applications that use UV photolysis for NDMA reduction. The Water Campus was built in the 1990s as a state-of-the-art wastewater treatment/recharge facility. The campus initially consisted of a 20-mgd water reclamation plant and a 10-mgd advanced water treatment (AWT) plant using microfiltration (MF) and reverse osmosis (RO) to treat the wastewater before groundwater injection. Upgrades were completed to the AWT facility starting in 2011 to increase the AWT capacity to 20 mgd. As part of this upgrade, a UV photolysis process was added to reduce NDMA concentration prior to reinjection. The general design criteria for the facility are presented in Table 10-4.

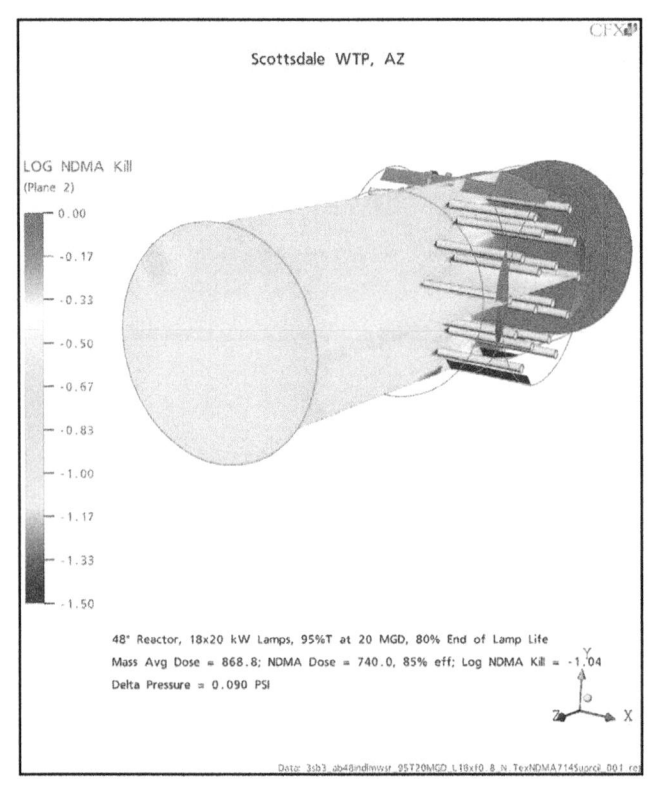

Figure 10-10   CFD-I model used to predict performance at the City of Scottsdale AWT and the UV reactor installed (courtesy of Calgon Carbon).

The UV photolysis process was designed without pilot testing. Sizing for the system was based on UV vendor modelling and a performance specification. The selected vendor was able to use computation fluid dynamics modelling coupled with intensity modelling (CFD-I) to predict the performance of the full-scale system (Figure 10-10). A 10-day full-scale performance test was required to verify the performance of the installed UV reactor. The performance testing showed that the UV reactor was able to consistently meet the design 1-log NDMA reduction under normal operating conditions (Figure 10-11).

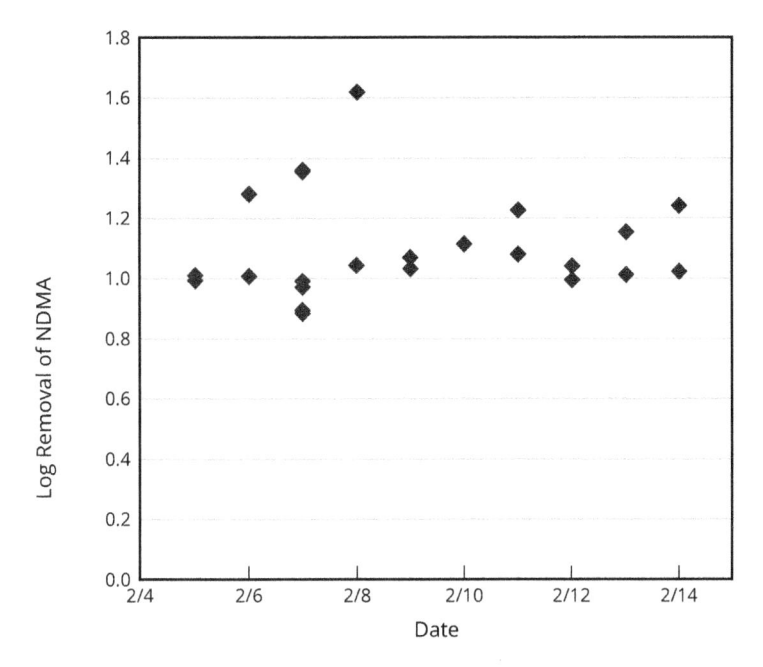

Figure 10-11 City of Scottsdale AWT performance test results (courtesy of Calgon Carbon).

11

Safety and Handling of AOT Equipment[1]

This chapter describes some of the safety issues associated with operation of AOT systems. The Occupational Safety and Health Administration (OSHA) issues regulations and guidance to support operator safety in the workplace, and there may also be specific safety requirements imposed by the governing agency. The following safety issues pertain to AOT systems:

- Electrical safety
- UV light exposure (UV AOTs)
- Burn safety
- Chemical safety (hydrogen peroxide, chlorine, ozone)
- Lamp breakage issues (if applicable): abrasions, cuts, and mercury exposure

ELECTRICAL SAFETY

When accessing the UV reactor, ozone generator, or power supply cabinets, manufacturer-recommended procedures should be followed. This includes, but is not limited to, disconnecting the main electrical supply and following proper lockout, tag-out procedures. The operator should allow for adequate time (e.g., at least 5 minutes) for the equipment to properly cool down and energy to dissipate before accessing the equipment. All federal, state, and local electrical codes (e.g., National Electric Code [NEC], OSHA), as well as manufacturer requirements, should be followed including the precautions below (USEPA 2006):

- Proper grounding
- Lockout, tag-out procedures
- Use of proper electrical insulators

---

[1] Some of this chapter has been adapted (with permission) from Bolton and Cotton (2008).

- Installation of safety cut-off switches
- Ground fault interrupt circuits

The design of the AOT system should also include proper grounding and insulation of electrical components, which will protect the equipment and operators during maintenance tasks.

UV LIGHT EXPOSURE

Typical UV reactors for AOT applications use closed-vessel reactors that limit the potential exposure to UV light. However, some existing installations use open-channel AOT reactors. Exposure to UV light can occur during reactor maintenance if proper precautions are not employed. There are no enforceable governmental standards for UV light exposure. However, for operator safety, the threshold limit value (TLV) for UV light exposure should be considered when establishing operations and maintenance procedures. TLVs can change periodically, and the guidebook *TLVs and BEIs Based on the Documentation of the Threshold Limit Values for Chemical Substances and Physical Agents and Biological Exposure Indices* (ACGIH 2012) should be referenced to determine the most recent values. The recommended TLVs are dependent on the wavelengths of the emitted UV light and the irradiance (mW cm^{-2}) and should be selected based on the installed equipment.

The UV lamps do not need to be energized during most operations and maintenance tasks. Also, the UV reactors should be designed with safety interlocks that turn off the UV lamps when the reactor is accessed. However, certain maintenance tasks (e.g., UV sensor checks) require the UV lamps to remain energized. Proper personal protective equipment (e.g., UV-resistant face shield, long-sleeve shirts, and gloves) should be worn at all times when the reactor is accessed while the UV lamps remain energized. If the UV reactors are provided with viewing ports, it should be verified that the viewing windows do not permit the transmittance of UV light. Warning signs should also be posted in any area where exposure to UV light is possible.

BURN SAFETY

UV lamps, depending on the lamp technology, may operate at up to 900°C. The UV lamps and sleeves should be allowed to cool properly before maintenance to minimize the risk of burns. Electrical equipment (e.g., ballasts) may also become hot during operation and should be evaluated prior to maintenance.

LAMP BREAK ISSUES

UV lamps pose two safety hazards if broken: (1) the lamps and sleeves are constructed of quartz that, on breaking, can pose a risk of serious cuts, and (2) UV lamps currently contain a small amount of mercury that can create an inhalation or contact hazard. Operators should be trained in proper cleanup procedures in case of a lamp break.

Lamp breaks are divided into two categories: off-line and online breaks. Off-line breaks occur when a lamp breaks during shipping, handling, storage, or maintenance. Off-line breaks can also occur when the lamp breaks in a UV reactor that is not in operation, or a UV lamp breaks during operation but the UV lamp sleeve does not break. An off-line break will expose water treatment plant (WTP) operators to mercury droplets and mercury vapor as well as sharp objects such as broken glass. An off-line lamp break would not release mercury into the distribution system.

Online lamp breaks occur when a lamp and lamp sleeve break while water is flowing through the UV reactor. These breaks are very rare, but they have the potential to release mercury into the water system. Potential causes of online lamp breaks include debris in the water, loss of water flow and temperature increases, pressure-related events, handling and maintenance errors, and UV reactor manufacturing problems.

Off-Line Lamp Breaks and Prevention Measures

In order to minimize the potential of an off-line lamp break, operators should be trained by the manufacturer to properly handle UV lamps and maintain the UV reactors. UV lamps should be stored horizontally in individual packaging and should not be stacked unpackaged or propped vertically against a wall.

On-line Lamp Breaks and Prevention Measures

On-line lamp breaks may pose similar hazards to the operators as off-line breaks as well as risks to water consumers if proper preventative measures are not practiced (USEPA 2006). Table 11-1 summarizes the potential causes of on-line lamp breaks and describes the preventive measures that can reduce each risk.

Mercury Regulatory Requirements

Federal and state regulations pertinent to mercury in drinking water, disposal of mercury as a hazardous waste, and mercury exposure in the workplace are summarized below.

- *Safe Drinking Water Act (SDWA)*—mercury in drinking water is regulated under SDWA. The USEPA set an enforceable maximum contaminant level (MCL) for inorganic mercury of 2 µg/L. Surface water utilities are required to conduct annual mercury compliance sampling unless mercury concentration trends are increasing or if an MCL has been exceeded, in which case, quarterly samples are required. MCL compliance is based on a running annual average. A single sample above the MCL would not constitute a violation, if the running annual average remains below the MCL. State regulatory agencies should be contacted to review any potential notification requirements for an online lamp break event.

- *UV Lamp Disposal*—mercury is listed as a hazardous waste under the Resource Conservation and Recovery Act (RCRA). USEPA requires federal, state, and

Table 11-1 Summary of on-line lamp break causes and prevention methods (USEPA 2006)

Potential Cause	Description	Preventive Measure
Debris	Physical impact of debris on lamp sleeves may cause lamp breaks.	Installation of screens, baffles, or low-velocity collection areas upstream of UV reactors or vertical installation of UV reactors will help prevent debris from entering the reactor.
Lamp orientation	Vertical installation relative to the ground may cause overheating and lamp breaks.	Install reactors with lamps oriented parallel to the ground to reduce differential heating.
Loss of water flow and temperature increases	Lamps may overheat and break. The temperature differential between stagnant water or air and flowing water (on resumption of flow) may cause lamp breaks.	Reactors should always be completely flooded and flowing during lamp operation. Temperature and flow sensors that are linked to an alarm and automatic shutoff system can be used to indicate irregular temperature or flow conditions.
Pressure-related events	Excessive positive or negative pressures may exceed lamp sleeve tolerances and break the lamp sleeve.	A surge analysis should be completed during design to determine the occurrence of water hammer.
		Pressure relief valves or other measures can be used to reduce pressure surges.
		Applicable pressure ranges should be specified for lamp sleeves.
Maintenance and handling errors	Improper handling or maintenance may compromise the integrity of the lamp sleeve and/or lamp.	Operators and maintenance staff should be trained by the manufacturer.
UV reactor manufacturing problems	Electrical surges can cause short-circuiting and lamp socket damage.	Adequate circuit breakers/GFI should be specified to prevent damage to the reactor.
	Applying power that exceeds design rating of lamps can cause lamps to burst from within.	Replacement lamps should be electrically compatible with reactor design.
	Misaligned or heat-fused cleaning mechanism may break or damage the lamp sleeve and lamp.	Operators and maintenance staff should perform routine inspection and maintenance according to manufacturers' recommendations.
	Thermally incompatible materials do not allow for expansion and contraction of lamp components under required temperature range.	Designers should specify temperature ranges likely to be encountered during shipping, storage, and operation of lamps to aid the manufacturer in the selection of thermally compatible materials.

Table 11-2 Health and safety standards for mercury compounds in air

Compound	OSHA PEL mg(Hg)/m^3	OSHA cPEL mg(Hg)/m^3	NIOSH IDLH mg(Hg)/m^3
Mercury compounds	Not reported	0.1	10
Organo alkyls containing mercury	0.01	0.04	2

local regulations be followed to ensure proper removal, transport, and disposal of hazardous waste.

- *Operator health and safety (exposure limits)*—the state and federal OSHA requirements should be consulted for mercury exposure to employees in the workplace, including personnel at water treatment plants. Exposure limits set by OSHA focus on exposure through inhalation. OSHA establishes permissible exposure limits (PEL) for compounds in air. A PEL is the time-weighted concentration not to be exceeded for an 8-hour workday during a 40-hour workweek. OSHA does not report a PEL for mercury but individual states may have enforceable standards (e.g., California OSHA PEL = 0.025 mg(Hg)/m^3. Another legal limit adopted by OSHA is a ceiling PEL (cPEL), which is defined as the concentration of mercury vapor that cannot be exceeded during any part of the workday. OSHA's cPEL for mercury is 0.1 mg(Hg)/m^3. PELs and cPELs are enforceable standards (USEPA 2006). An immediately dangerous to life or health (IDLH) concentration for mercury was published by the National Institute for Occupational Safety and Health (NIOSH). IDLH represents the maximum concentrations that one could escape within 30 minutes without symptoms of impairment or irreversible health effects. Table 11-2 summarizes the federal health and safety standards for mercury compound at the workplace.

Proper and prompt lamp break response can typically maintain mercury vapor concentrations below the PELs, cPELs, and IDLHs. If proper cleanup is not followed, the mercury vapor concentration may exceed the limits in areas where mercury can collect (e.g., a drained UV reactor after an online break) (USEPA 2006). The proper cleanup procedures should be outlined in a mercury release response plan.

MERCURY RELEASE RESPONSE

Response plans for off-line and online lamp breaks should be developed. Potential items to be included in these response plans are described in the following section. Mercury and materials used during the cleanup procedure are regulated as hazardous wastes and should be disposed of properly.

Off-Line Lamp Break Response

Proper storage and handling of UV lamps helps to minimize the risk of an off-line lamp break. In the event that a lamp break occurs, a lamp break response plan for containing and cleaning up off-line mercury spills should be available to ensure proper safety procedures are followed.

The size of the mercury release determines the appropriate response to the spill. The USEPA Office of Emergency and Remedial Response recommends that "[in] the event of a large mercury spill (more than a broken thermometer's worth), immediately evacuate everyone from the area, seal off the area as well as possible, and call your local authorities for assistance" (USEPA 1997). If a lamp break results in a small spill, the mercury can be contained and cleaned up with a commercially available mercury spill kit. Small spills are defined as the amount of mercury in a broken thermometer, or less than 2.25 g (USEPA 1992, USEPA 1997). Lamp breaks typically result in a small spill based on typical UV lamps containing between 0.005 and 2 g of mercury (USEPA 2006).

Online Lamp Break Response

An online lamp break is unlikely; however, an online lamp break response plan should be developed prior to operation so that proper response and cleanup measures are followed if an online break occurs. The following is a list of USEPA suggested components of a lamp break response plan (USEPA 2006). The governing agency should also be consulted when developing the response plans.

- Identification of a lamp break
- Site-specific containment measures
- Mercury sampling and compliance monitoring
- Site-specific cleanup procedures
- Reporting and public notification

There are three major steps to responding to a UV lamp break event

- Prevention
- Cleanup
- Investigation

Prevention

Online lamp breaks are rare incidents, and all UV AOT facilities should be designed to minimize conditions that could cause lamp and sleeve breaks. Protection can include upstream treatments (e.g., filtration) and proper controls and operations to limit the potential for high flow events. Many UV reactors are also programmed with alarms that will indicate potential lamp break events and trigger the specific UV reactor train to be isolated and shut down.

Cleanup

Three cleanup response options can be evaluated as possible actions following an online lamp break event, including

- **Containment**—UV reactor trains have isolation valves that will automatically close if a UV lamp break alarm is generated. However, the slug of water containing mercury will likely have passed these isolation valves faster than the valves can close and therefore cannot be captured within the UV train. Depending on the location of isolation valves downstream from the UV AOT facility, downstream valves could be closed to isolate the slug of water containing mercury. Once isolated, the water would have to be drained and treated for disposal. The option for containment may be limited depending on the size of the facility and location of isolation valves.

- **Diversion**—diversion of water following an online lamp break event would require closing a valve downstream of the facility to isolate the distribution system and opening a second valve to divert water to a containment area. The location of the downstream valves would have to be located far enough away to allow adequate time to open and close the valves prior to the slug of water reaching the diversion point. As with the containment option, treatment and ultimate disposal of the water should be considered.

- **Monitoring**—containment or diversion of the mercury containing water is often not possible due to the required response time to open or close valves and the multiple pathways that the water can often travel after leaving a plant. For these cases, monitoring of mercury concentrations is typically the selected approach. Upfront modeling can also be completed to predict what the mercury concentration would be if an event occurred. For most instances, concentrations will be below the mercury MCL, especially for facilities that have clearwells downstream of the UV reactors. Table 11-3 includes potential monitoring locations.

Investigation

An investigation into the cause of the lamp break will help to prevent escalation and/or additional events. Investigation of a lamp break occurrence involves identification of the cause of the break and modifcations to the control strategies or physical modifications to the facility necessary to prevent future breaks.

CHEMICAL SAFETY

AOT systems typically include some type of chemical addition, which may include hydrogen peroxide, chlorine, or ozone. Each of these chemicals has safety concerns, and designers and operators of the facility should be aware of the concerns in order to minimize the risk of potential accidents or harm.

Table 11-3 Mercury sampling locations (USEPA 2006)

Media	Location	Purpose
Water	• Reactor drain • Piping downstream of the UV reactor, including the distribution system entry point at a minimum • Low velocity areas, such as clearwells	• Assess the extent of mercury contamination and identify areas requiring cleanup.
Air*	• Reactor or other locations where mercury vapor may collect • Ambient air	• Assess whether it is safe to access mercury-contaminated equipment and piping for cleanup. The UV reactor interior may be accessible through an air vent. • Assess whether adequate ventilation is provided to safely proceed with mercury cleanup.

* Methods for air sampling are available from the OSHA at http://www.osha.gov/dts/sltc/methods/inorganic/id140/id140.html.

Hydrogen Peroxide

Hydrogen peroxide is typically delivered at 35 percent or 50 percent solutions. Hydrogen peroxide is a strong oxidant and should not be allowed to come in contact with organic materials or become contaminated due to rapid degradation. Contact with the chemical or vapors can cause skin or mucous membrane irritation. Storage facilities must provide ample ventilation and water must be readily accessible for flushing in the case of a spill. Proper personal protective equipment should be worn at all times when handling hydrogen peroxide.

Chlorine

Chlorine is most often applied as sodium hypochlorite and is typically delivered as 12.5 percent solution. Sodium hypochlorite should be kept isolated from acids or other low-pH solutions and ammonia solutions to avoid generation of chlorine gas. Chlorine poses similar risks as hydrogen peroxide in that contact or inhalation of the chemical can be severely irritating to the skin and mucous membranes. Storage and delivery should be designed to allow for ventilation and easy access to water in the case of a spill. Proper personal protective equipment should be worn at all times when handling sodium hypochlorite.

Ozone

Ozone off-gassing or leaks from AOT systems in the air can pose a health risk to workers in the area. Exposure to ozone can cause respiratory system irritation including coughing, throat irritation, discomfort, and chest tightness. Adverse health effects from short-term exposure to ozone are reversible by discontinuing exposure to ozone. Buildings housing ozone systems should be equipped with adequate ventilation and ambient air monitors

Table 11-4 Health and safety standards for ozone in air

Compound	OSHA PEL (ppm)	OSHA cPEL (ppm)	NIOSH IDLH (ppm)
Ozone	0.1 ppm	Not available	5 ppm

to monitor ozone concentration. The OSHA PEL for ozone is 0.1 ppm based on a time weighted average. The NIOSH also has an IDLH concentration of 5 ppm (Table 11-4). Any ozone-based AOT should include ambient air ozone analyzers for operator safety. Ozone generation systems should have interlocks to shut down the ozone generation upon detection of a leak.

Many ozone systems use liquid oxygen (LOX) as the oxygen source for ozone generation. LOX itself is not a flammable chemical but it can lower the ignition point of other materials and is a strong oxidizing agent. Therefore, organic materials will rapidly burn in the presence of LOX; so it is recommended that LOX storage be placed at least 50 ft from the use of any organic materials. LOX is stored cryogenically; therefore, there is a frostbite hazard if exposed to the liquid or extended exposure to LOX piping. Finally, an oxygen leak could create an oxygen enriched environment (>25 percent), which can cause lung damage upon extended exposure. Oxygen monitors should be included in enclosed spaces where oxygen piping is present.

REFERENCES

ACGIH. 2006. *TLVs and BEIs Based on the Documentation of the Threshold Limit Values for Chemical Substances and Physical Agents and Biological Exposure Indices.* Cincinnati, OH: American Conference of Governmental Industrial Hygienists.

Bolton, J.R., and C.C. Cotton. 2008. *The Ultraviolet Disinfection Handbook.* Denver, CO: American Water Works Association.

USEPA. 1992. *Characterization of Products Containing Mercury in Municipal Solid Waste in the United States, 1970–2000.* Washington, DC: Office of Solid Waste.

USEPA. 1997. *Mercury—Emergency Spill and Release Facts.* EPA 540-K-97-004, OSWER 9378.0-10FS, PB97-963405. Washington, DC: Office of Emergency and Remedial Response, US Environmental Protection Agency.

USEPA. 2006. *Ultraviolet Disinfection Guidance Manual.* Washington, DC: Office of Water.

12

Considerations for a Water Utility Manager[1]

This chapter is designed to provide guidance for a water utility manager or water quality manager whose utility is interested in exploring the possible installation of an AOT. Several decisions need to be made at each stage of the process.

USING AN ENGINEERING CONSULTING FIRM

Most large water utilities will procure an engineering consultant to design an AOT facility. However, smaller utilities (less than 1 mgd) may prefer an in-house design, depending on their financial situation. The advantage of using an engineering consulting firm is that the utility can use the prior experience of the firm concerning AOT equipment design and procurement, facility design, safety, and operations and maintenance issues. The disadvantage is that it can potentially increase the cost of the project and may not provide knowledge transfer to in-house staff.

The advantage of doing an in-house design is that the utility can deal directly with equipment vendors and design the UV facility using the utility's engineering staff. There is more control over the process, and additional knowledge may be gained by the engineering staff; however, there must be considerable knowledge of AOTs and the various types of equipment. Also, there must be engineers or staff available who are capable of managing the total project. AOT equipment does not have validation reports as are available with UV disinfection applications to prove equipment will work as designed. This places more importance on the selection of a knowledgeable equipment manufacturer that will provide performance guarantees for the system. Also, additional importance is placed on a proper understanding of the design and operating conditions for the facility.

INFORMATION NEEDS

If AOT is being considered, early collection of data will help refine the design and equipment selection and possibly reduce design, construction, and operation costs. The relevant information can be summarized as the following:

---

1 Some of this chapter has been adapted (with permission) from Bolton and Cotton (2008).

- Flow rate (mgd or ML/d) variability on which to base the design flow
- Complete description of the facility as it currently operates and any planned modifications, including the following water quality data at a minimum:
  - Target contaminant
  - TOC
  - Alkalinity
  - pH
  - UVT at 254 nm and UV scans (UV only)
  - Nitrate (medium-pressure UV only)
  - Possible foulants (e.g., calcium, alkalinity, hardness, iron, manganese, pH, ORP) (UV only)
  - Bromide (ozone or UV/chlorine only)
  - Temperature
- Seasonality of target contaminant(s) coupled with flows and water quality
- Hydraulic profile throughout existing facilities and allowable head loss
- Potential locations in the treatment train of the AOT system—for example, upstream of filtration or combined filter effluent location
- Plans to build a new facility or retrofit the equipment into the existing plant; if a retrofit, evaluate location in the plant and the space available for installation

QUESTIONS FOR A POTENTIAL ENGINEERING CONSULTING FIRM

If an engineering consulting firm is selected, several important questions should be addressed. The possible engineering consulting firms should be approached either with a request for qualifications (RFQ) or proposals (RFP), with the procurement details that meet the local requirements. Local procurement codes will dictate most of the submittal requirements; however, potential specific AOT elements to be considered in the solicitation are

- What experience does the firm have in designing and constructing AOT facilities locally and nationally?
- What is the previous work experience with the utility?
- What is the firm's experience with testing AOT systems (bench scale, pilot scale, and full-scale)?
- Provide references regarding previous AOT projects.

QUESTIONS FOR A POTENTIAL EQUIPMENT MANUFACTURER

If an in-house design has been chosen, the utility will deal directly with AOT equipment manufacturers. The utility should develop an AOT equipment specification that will be provided to the potential AOT manufacturers. The specification could include the items described in Table 12-1.

A list of qualified AOT vendors should be prepared using criteria such as having operational AOT facilities of similar size to that of the utility, references for operating facilities, results showing performance of installed equipment.

Table 12-1   Possible content for AOT equipment specifications (adapted from USEPA 2006)

Item	Specification Content
Flow rate	Maximum, minimum, and average flow rates should be clearly identified. The minimum flow rate may be important to avoid overheating with UV reactors.
Target contaminant(s) and log inactivation	The log inactivation for the target contaminant(s).
Water quality and environment	The following water quality criteria should be included: • Influent temperature • TOC • UV transmittance at 254 nm • UVT scan from 200–300 nm (MP reactors only) • Total alkalinity • Total hardness • Bromide • pH • Iron • Calcium • Manganese • ORP • Nitrate For some parameters, a design range may be most appropriate.
Operating flow and UVT matrix (UV only)	Appropriate matrix of paired flow and UVT values based on flow and UVT data (if applicable).
Operating flow, water quality, and target log reduction	Appropriate matrix of flow, water quality, and log reduction values based on seasonal variations in target contaminant (if applicable).
Operating pressure	The expected operating pressures, including the maximum and minimum operating pressure.
Disinfection credit required	Identify equipment that will also be used for disinfection credit. Include target pathogens and log inactivation.
Maximum oxidant dose	Maximum oxidant dose, if desired, to protect equipment and minimize quenching requirements.

Table continues next page

Table 12-1 Possible content for AOT equipment specifications (adapted from USEPA 2006) (continued)

Item	Specification Content
UV sensors (UV only)	A germicidal spectral response should be specified, especially if disinfection is also desired. A minimum of one UV sensor should be specified per UV reactor. Low wavelength sensors for MP UV reactors may be beneficial once the sensors become available. Reference UV sensors should be calibrated against a traceable standard. For example, the following standards are currently being used by UV manufacturers for disinfection applications: • National Physical Laboratory (NPL) • National Institute of Standards and Technology (NIST) • Deutsche Vereinigung des Gas- und Wasserfaches (DVGW) • Österreichisches Normungsinstitut (ÖNORM)
Redundancy	Specify required number of redundant/standby units. Redundancy requirements could include the following, as applicable: • UV reactors • Online analyzers • Ozone generators • Chemical feed pumps • Ozone destruct units • Cooling water pumps • Ozone injection system components • Nitrogen boost units • Sidestream injection pumps
Hydraulics	The following hydraulic information should be specified: • Maximum system pressure • Maximum allowable head loss • Special surge conditions that may be experienced • Hydraulic constraints based on site-specific conditions (e.g., upstream and downstream straight pipe lengths)
Size/location constraints	Any size constraints or restrictions on the location (e.g., space constraints with individual filter effluent installation).
Validation (disinfection credit with UV only)	The range of operating conditions (e.g., flow, UVT) that must be included in the validation testing, and submittal of a validation report (40 CFR 141.720) should be required. The validation testing should be completed in accordance with the procedures and data analysis described in the UVDGM.
Monitoring strategy	A description of the preferred E_{EO} or dose-monitoring strategy for the UV reactors. CT monitoring strategy for ozone systems.
Operating approach	A description of the intended operating approach for the AOT system.
Economic and noneconomic factors	The necessary information to thoroughly evaluate the equipment based on the utility's specific goals. As appropriate, this information may include both economic (e.g., energy use, chemical use) and noneconomic (e.g., future expansion, manufacturer experience) factors.

Table continues next page

Table 12-1 Possible content for AOT equipment specifications (adapted from USEPA 2006) (continued)

Item	Specification Content
Lamp sleeves (UV only)	Lamp sleeves should be annealed to minimize internal stress.
Ozone and ancillary equipment (ozone only)	Type of ozone generators, oxygen source (LOX versus air feed), LOX storage and feed, ozone injection system, ozone destruct units, and nitrogen boost.
Performance guarantee	The equipment provided should meet the performance requirements stated in the specification for an identified period and/or operating conditions during on-site performance testing. The following specific performance criteria may be included: • Allowable head loss at each design flow rate • Estimated power consumption under the design operating conditions • Oxidant dose under the design conditions • Sensitivity of equipment to variations in voltage or current • Analyzer performance compared to specification (e.g., ozone, UVT, or UV sensor)
Warranties	A physical equipment guarantee and component guarantees should be specified. The specific requirements of these guarantees will be at the discretion of the utility and engineer. Significant variation from common commercial standards should be discussed with the manufacturer. To limit the manufacturer's liability, the component guarantees may be prorated after a specified number of operating hours.
Analyzers	During operation, the difference between the online analyzer measurements (e.g., ozone or UVT) and grab samples should be measured to verify calibration and accuracy of the analyzers. Analyzers that cannot maintain calibration should be replaced by the manufacturer.
Spare parts	Number of spare parts that should be included as part of the base bid price. Specification may also include unit costs for additional spare parts.
Hydrogen peroxide quenching	Equipment required for quenching, if required and included as part of the AOT system specification. Specification should include performance guarantee for the quenching system as well as AOT system.

GOVERNMENT REGULATIONS

AOT systems are not as well regulated as disinfection systems, and there is no equivalent to the USEPA UVDGM (USEPA 2006) or *Alternative Disinfectants and Oxidants Guidance Manual* (USEPA 1999) for AOT systems. However, the state or provincial regulator should be contacted early in the planning process and then strategically throughout the project to ensure a smooth permitting process for the AOT facility.

REFERENCES

Bolton, J.R., and C.C. Cotton. 2008. *The Ultraviolet Disinfection Handbook*. Denver, CO: American Water Works Association.

USEPA. 1999. *Alternative Disinfectants and Oxidants Guidance Manual.* Washington, DC: Office of Water.

USEPA. 2006. *Ultraviolet Disinfection Guidance Manual.* Washington, DC: Office of Water.

Appendix A

Terms, Units, Symbols and Definitions

This appendix contains definitions of the terms, units, symbols, and definitions used in this book. The definitions marked with the † symbol have been obtained from http://www.pbs.org/faithandreason/gengloss/index-frame.html. Symbols follow the term and the units are in square brackets; [1] means no units.

UNITS AND PHYSICAL CONSTANTS

All of the units used in this book conform to the Système International (SI). An excellent reference for terms and nomenclature in physical chemistry is the International Union of Pure and Applied Chemistry (IUPAC) "Green Book" (Cohen et al., 2011). Some of the physical constants of interest are listed in Table A-1.

Another important reference is the Glossary of Terms used in *Photochemistry* (Braslavsky 2007).

Note that there are two kinds of wavelength dependent parameters: (1) those that have functional dependence on wavelength {e.g., the molar absorption coefficient $[\varepsilon(\lambda)]$}, and (2) those that are the result of differentiation with respect to the wavelength [e.g., the spectral irradiance (E_λ)]. The wavelength dependence of the former will be designated with (λ) at the end of the symbol, and the wavelength dependence of the latter will be designated with a subscript (e.g., E_λ). One can always tell which parameters are of type 2—they will have nm^{-1} as part of the units.

Table A-1 Physical constants of interest in ultraviolet technologies

Constant	Symbol	Value	Units
Speed of light	c	2.99792458×10^8	m s^{-1}
Charge on electron	e	$1.60217733 \times 10^{-19}$	C
Planck constant	h	$6.6260755 \times 10^{-34}$	J s
Boltzmann constant	k	1.380658×10^{-23}	J K^{-1}
Avogadro number	N_A	6.0221367×10^{23}	mol^{-1}

The following is an alphabetical list of the terms (with definitions) as used in this book.

TERMS AND DEFINITIONS

absorbance, $A(\lambda)$ [1] – decadic logarithm ($\log_{10}$) of the ratio of the incident to the transmitted irradiance as a beam with a narrow set of wavelengths centered on the wavelength λ passes through a medium over a path length l. The absorbance is related to the transmittance $T(\lambda)$ by the expression $A(\lambda) = -\log[T(\lambda)]$.

absorption – physical process of the removal of photons from a beam as it passes through a medium containing absorbing substances.

absorption coefficient (decadic), $a(\lambda)$ [m^{-1} or cm^{-1}] – the absorbance A divided by the path length l.

absorption coefficient (naperian), $\alpha(\lambda)$ [m^{-1} or cm^{-1}] – the absorption coefficient (naperian) is used when the decay of irradiance in a medium is expressed as an exponential. Note that $\alpha(\lambda) = \ln(10)a(\lambda)$.

absorption spectrum – a plot of the absorbance $A(\lambda)$ versus wavelength λ.

action spectrum – plot of a relative biological or chemical photoresponse (= Δy) per number of incident (prior to absorption) photons, vs. wavelength, or energy of radiation, or frequency or wavenumber. This form of presentation is frequently used in the studies of biological or solid-state systems, where the nature of the absorbing species is unknown. It is advisable to ensure that the fluence dependence of the photoresponse is the same (e.g., linear) for all the wavelengths studied.

> Note 1: The action spectrum is sometimes called the *spectral responsivity* or *sensitivity spectrum*. The precise action spectrum is a plot of the spectral (photon or quantum) effectiveness. By contrast, a plot of the biological or chemical change or response per absorbed photon (quantum efficiency) vs. wavelength is the efficiency spectrum.

> Note 2: In cases where the fluence dependence of the photoresponse is not linear (as is often the case in biological photoresponses), a plot of the photoresponse vs. fluence should be made at several wavelengths and a standard response should be chosen (e.g., two-log reduction). A plot of the inverse of the 'standard response' level vs. wavelength is then the action spectrum of the photoresponse.

advanced oxidation technologies (AOTs) – Technologies that generate highly reactive intermediates (e.g., hydroxyl radicals) to oxidize and degrade organic contaminants in water.

CFD (computational fluid dynamics) – a mathematical model that calculates the paths of microorganisms as they flow through a pipe or UV reactor taking account of the laws of fluid dynamics.

collimated beam apparatus – a bench scale apparatus consisting of a UV lamp in an enclosure that allows a narrow beam to be collimated, either by apertures or a collimating tube. A Petri dish containing a suspension of microorganisms is placed under the beam. This is also called *quasi collimated beam apparatus*, because the beam is not exactly parallel.

concentration, c [M] – the amount (moles) of a substance in solution per liter of solution.

DNA (deoxyribonucleic acid) – a double-stranded helix of nucleotides that carries the genetic information of a cell. It encodes the information for the synthesis of proteins and is able to self-replicate.[‡]

einstein – one mole (6.02214×10^{23}) of photons.

electromagnetic radiation (J) – energy transmitted at the speed of light (in a vacuum) and characterized by its division into photons, which have an energy inversely proportional to the wavelength. Often the abbreviated term *radiation* is used.

electrical energy dose, EED [kWh m^{-3}] – total electrical energy consumed in driving an AOT per m^3 of solution.

electrical energy per mass, E_{EM} [kWh kg^{-1}] – total electrical energy (kWh) required to degrade 1 kg of a contaminant in an AOT solution.

electrical energy per order, E_{EO} [kWh order^{-1} m^3] – total electrical energy required to reduce the concentration of a contaminant by one order of magnitude (90%) in driving an AOT per m^3 of solution.

fluence, F_o [J m^{-2}] – total radiant energy Q incident from all directions onto a small sphere divided by the cross-sectional area of that sphere. Fluence is often referred to as UV dose. The units mJ cm^{-2} are commonly used; 1 mJ cm^{-2} = 10 J m^{-2}.

fluence rate, E_o [W m^{-2} or mW cm^{-2}] (see also *irradiance*) – the total radiant power incident from all directions onto an infinitesimally small sphere of cross-sectional area dA, divided by dA (see Figure A-1b). Although the unit of W m^{-2} is in use throughout most of the world, the unit mW cm^{-2} is widely used in North America [1 mW cm^{-2} = 10 W m^{-2}].

fluorescence – emission of photons when molecules return to their ground state from excited singlet states of the molecules. The fluorescence occurs at longer wavelengths than that of the exciting light.

high-energy radiation – particles (alpha or beta) or electromagnetic radiation (gamma rays) emitted by radioactive substances or high-energy electron beams. High-energy radiation is characterized by causing ionization when absorbed in a substance. This generates high-energy electrons and radicals that cause nonspecific damage to the substance.

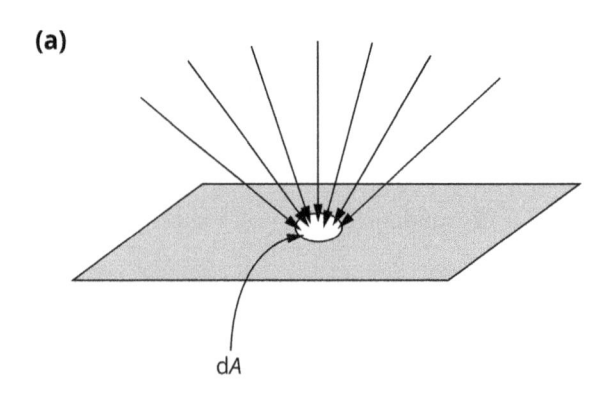

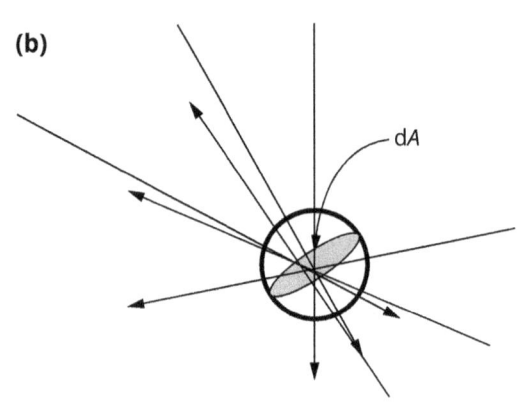

Figure A-1 Illustration of the concepts of irradiance and fluence rate: (a) *irradiance* onto a surface; (b) *fluence rate* through an infinitesimally small sphere of cross-sectional area d*A*

intensity, I [mW cm^{-2}] – this term is used extensively in the USEPA Ultraviolet Disinfection Guidance Manual (UVDGM) (USEPA, 2006). It generally has the same meaning as *irradiance*, although in some cases it means *fluence rate*. The UVDGM does not define these terms adequately. Note that the unit of W m^{-2} is in use throughout most of the world, but the unit mW cm^{-2} is widely used in North America [1 mJ cm^{-2} = 10 J m^{-2}].

irradiance, E [W m^{-2} or mW cm^{-2}] – the total radiant power *incident* from all incoming directions *on* an infinitesimal element of *surface* of area dS containing the point under consideration divided by dS (see Figure A-1a). Although the unit of W m^{-2} is in use throughout most of the world, the unit mW cm^{-2} is widely used in North America [1 mW cm^{-2} = 10 W m^{-2}]. The UVDGM (USEPA, 2006) uses the symbol I for irradiance and calls it intensity.

light – electromagnetic radiation in the UV, visible or IR regions of the spectrum. Note that physicists restrict the definition of light to the visible portion of the spectrum.

LP lamps (low pressure lamps) – UV lamps containing a very low amount of mercury, such that the mercury pressure in the gas phase is a few Pascals.

LPHO lamps (low pressure high output lamps) – UV lamps with either a large diameter or containing an amalgam of mercury with another element (e.g., gallium). The output of LPHO lamps is 2–3 times that of LP for the same length.

molar absorption coefficient, $\varepsilon(\lambda)$ [M^{-1} cm^{-1}] – the absorbance of a solution of concentration 1 M and with a path length of 1 cm. Note that the molar absorption coefficient is a function of wavelength.

MP lamps (medium pressure lamps) – UV lamps containing a moderate amount of mercury, such that the mercury pressure in the gas phase is around 100 Pascals (1 atm). The output of MP is 20 – 50 times that of LP lamps for the same length, and the spectral emission is much broader.

path length, l [m or cm] – the distance over which a beam of UV light passes through a medium.

phosphorescence – photon emission from long-lived excited triplet states of a molecule as the excited state returns to the ground state.

photochemistry – chemistry induced by the absorption of light.

photon – the fundamental 'particle' of electromagnetic radiation (e.g., light). A photon has no mass but has an energy inversely proportional to the wavelength.

photon fluence, $F_{p,o}$ [einstein m^{-2}] – total number of moles of photons (einsteins) incident from all directions on a small sphere over time divided by the cross-sectional area of the sphere.

photon fluence rate, $E_{p,o}$ [einstein m^{-2} s^{-1}] – rate of *photon fluence*. Total number of moles of photons (einsteins) incident from all directions on a small sphere per time interval divided by the cross-sectional area of the sphere.

photon flux, q_p [einstein s^{-1}] – number of moles of photons (einsteins) passing per time interval.

photon irradiance, E_p [einstein m^{-2} s^{-1}] – number of moles of photons (einsteins) per time interval (photon flux) q_p incident from all upward directions on a small element of surface containing the point under consideration divided by the area of the element.

quantum yield, Φ [1]

Number of defined events occurring per photon absorbed by the system. For a photochemical reaction

$$\Phi = \frac{\text{moles of reactant consumed or product formed}}{\text{einsteins absorbed}}$$

Strictly, the term quantum yield applies only for monochromatic excitation.

radiant emittance, $M(\lambda)$ [W m^{-2}] (also called *excitance*) of a source – the radiant power emitted in all outward-bound directions from an infinitesimal area dA on the surface of the source. The radiant emittance is a measure of the *brightness* of a source.

radiant energy, Q [J] – the total amount of radiant energy emitted from a source over a given period of time.

radiant exposure, H [J m^{-2} or mJ cm^{-2}] – the total radiant energy *incident* from all upward directions on an infinitesimal element of *surface* of area dS containing the point under consideration divided by dS (see Figure A-1a). Although the unit of J m^{-2} is in use throughout most of the world, the unit mJ cm^{-2} is widely used in North America [1 mJ cm^{-2} = 10 J m^{-2}].

radiant power, P_Φ [W] – the radiant power of a source is the rate of radiant energy emission or the total radiant power emitted in all directions by a light source.

$$P_\Phi = \frac{dQ}{dT} \qquad \text{(Eq. A-1)}$$

For example, the radiant power of the Sun is 3.842 × 10^{26} W. In theory, P_Φ should include all wavelengths emitted by the source; however, P_Φ is usually restricted to the wavelength range of interest for photochemistry. For example, if a light source is being used for ultraviolet photochemistry, P_Φ would be specified for emission in the 200–400 nm ultraviolet range.

radiant power efficiency, η [1] – the *radiant power efficiency* of a lamp is defined as

$$\eta = \frac{P_\Phi}{P_E} \qquad \text{(Eq. A-2)}$$

where P_E is the input electrical power (W) from the wall to run the lamp and its power supply. Sometimes P_E is the electrical power across the lamp.

radiation – see *electromagnetic radiation* and *high energy radiation*.

RED (reduction equivalent dose) – the UV dose (fluence) delivered by a UV reactor as determined by a biodosimetry test.

REF (reduction equivalent fluence) – the fluence (UV dose) delivered by a UV reactor as determined by a biodosimetry test.

refractive index (for a given medium), n – ratio of the speed of light in a vacuum to that in a given medium.

spectral fluence, $F_{\lambda,o}$ [J m^{-2} nm^{-1}][1] – derivative of fluence, F_o, with respect to wavelength λ.

spectral fluence rate, $E_{\lambda,o}$ [W m^{-2} nm^{-1}] – derivative of fluence rate E_o with respect to wavelength λ.

spectral irradiance, E_λ [W m^{-2} nm^{-1}] – derivative of irradiance E with respect to wavelength λ.

spectral first-order rate constant, $k_{1,\lambda}$ [s^{-1} nm^{-1}] – derivative of the first-order rate constant k_1 with respect to wavelength λ.

spectral photon flux, $q_{p,\lambda}$ [einstein s^{-1} nm^{-1}] – derivative of the photon flux q_p with respect to wavelength λ.

spectral photon fluence rate, $E_{p,o,\lambda}$ [einstein s^{-1} m^{-2} nm^{-1}] – derivative of the photon fluence rate $E_{p,o}$ with respect to wavelength λ.

spectral photon irradiance, $E_{p,\lambda}$ [einstein s^{-1} m^{-2} nm^{-1}] – derivative of the photon irradiance E_p with respect to wavelength λ.

transmittance, $T(\lambda)$ [1] – ratio of the transmitted irradiance to that of the incident irradiance as a beam passes through a medium over a path length l. The transmittance is related to the absorbance $A(\lambda)$ by the expression $T(\lambda) = 10^{-A(\lambda)}$.

transmittance spectrum – a plot of the transmittance $T(\lambda)$ versus wavelength λ.

UV dose (fluence), F_o [J m^{-2} or mJ cm^{-2}] (see *fluence*) The UVDGM (USEPA, 2006) uses the symbol, D for UV dose (fluence).

UVT (UV transmittance), [%] – percent transmittance of a beam of UV light as it passes through a medium over a path length of 1 cm.

wavelength, λ [nm] – the distance between successive crests of a wave, such as a sound wave or electromagnetic wave.

REFERENCES

Braslavsky, S.E., 2007. Glossary of Terms used in Photochemistry. 3rd ed. *Pure Appl. Chem.* 79(3): 292-465 http://www.iupac.org/publications/pac/2007/7903/7903x0293.html).

Cohen, E.R., T. Cvitaš, J.G. Frey, B. Holmström, K. Kuchitsu, R. Manquardt, I. Mills, F. Pavese, M. Quack, J. Stohner, H.I. Strauss, M. Takami and A.J. Thor. 2011. IUPAC Green Book: *Quantities, Units and Symbols in Physical Chemistry*, 3rd Ed., Cambridge, UK: RSC Publishing.

USEPA, 2006. *Ultraviolet Disinfection Guidance Manual for the Final Long Term 2 Enhanced Surface Water Treatment Rule*. Washington, DC: US Environmental Protection Agency. http://www.epa.gov/safewater/disinfection/lt2/pdfs/guide_lt2_uvguidance.pdf.

---

1 Strictly speaking, the SI units would be J m^{-3}; however, the units of nm are retained for clarity. Similar statements apply to the other spectral quantities.

Appendix B

Rate Constants and Quantum Yields

Table B-1 ·OH radical Rate Constants

Reagent	Rate Constant ($M^{-1} sec^{-1}$)	Notes	Reference
·OH	1.1×10^{10}		Buxton et al. (1988)
HO_2·	6×10^9		Buxton et al. (1988)
O_2^-·	8×10^9		Buxton et al. (1988)
O_3	1.1×10^8		Buxton et al. (1988)
H_2O_2	2.7×10^7	$pK_a = 11.8$	Buxton et al. (1988)
HO_2^-	7.5×10^9		Christensen et al. (1982)
HCO_3^-	8.5×10^6	$pK_a = 6.4$	Buxton et al. (1988)
CO_3^{2-}	3.9×10^8		Buxton et al. (1988)
$H_2PO_4^-$	$\sim 2 \times 10^4$	$pK_a = 5.7$	Maruthamuthu and Neta (1978)
HPO_4^{2-}	1.5×10^5		Maruthamuthu and Neta (1978)
PO_4^{3-}	$< 1.0 \times 10^7$		Buxton et al. (1988)
HSO_4^-	4.7×10^5		Jiang et al. (1992)
Cl^-	3.0×10^9	pH < 2	Grigor'ev et al. (1987)
HOCl	8.5×10^4		Watts and Linden (2007)
OCl^-	8.8×10^9		Buxton et al. (1988)
Br^-	1.1×10^{10}	pH ~ 1	Zehavi and Rabani (1972)
I^-	1.1×10^{10}	pH = 7	Buxton et al. (1988)
NO_2^-	1.0×10^{10}		Buxton et al. (1988)
CH_3OH	8.3×10^8		Motohashi and Saito (1993)
$H_2C(OH)_2$	2.6×10^7	(hydrated HCHO)	Chin and Wine (1994)
HCOOH	1.1×10^8		Chin and Wine (1994)
$HCOO^-$	3.1×10^9		Chin and Wine (1994)
CH_3CN	2.2×10^7		Buxton et al. (1988)

Table continues next page

Table B-1 ·OH radical Rate Constants (continued)

Reagent	Rate Constant (M^{-1} sec^{-1})	Notes	Reference
C$_2$H$_5$OH	5.9×10^9		Motohashi and Saito (1993)
CH$_3$CHO	3.6×10^9		Schuchmann and von Sonntag (1988)
CH$_3$COCH$_3$	1.1×10^8		Buxton et al. (1988)
CH$_3$COOH	1.6×10^7		Buxton et al. (1988)
CH$_3$COO$^-$	8.5×10^7		Buxton et al. (1988)
CH$_2$Cl$_2$	9×10^7		Haag and Yao (1992)
CHCl$_3$	5×10^7		Haag and Yao (1992)
Cl$_2$C=CCl$_2$	2.6×10^9		Buxton et al. (1988)
Cl$_2$C=CHCl	4.2×10^9		Buxton et al. (1988)
CH$_3$CH$_2$CH$_2$OH	2.8×10^9		Buxton et al. (1988)
CH$_3$CHOHCH$_3$	1.9×10^9		Buxton et al. (1988)
(C$_2$H$_5$)O(C$_2$H$_5$)	3.6×10^9		Buxton et al. (1988)
(CH$_3$)$_3$OCH$_3$	1.6×10^9		Eibenberger (1980)
CH$_3$SOCH$_3$	6.6×10^9		Buxton et al. (1988)
(CH$_3$)$_2$NNO	3.3×10^8		Wink et al. (1991)
H$_2$C$_2$O$_4$	1.4×10^6		Getoff et al. (1971)
HC$_2$O$_4^-$	4.7×10^7		Getoff et al. (1971)
C$_2$O$_4^{2-}$	7.7×10^6		Buxton et al. (1988)
1,4-Dioxane	2.8×10^9		Buxton et al. (1988)
Benzene	7.9×10^9		Ashton et al. (1995)
Toluene	5.1×10^9		Roder et al. (1990)
p-Xylene	7.0×10^9		Sehested et al. (1975)
m-Xylene	7.5×10^9		Sehested et al. (1975)
o-Xylene	6.7×10^9		Sehested et al. (1975)
Ethylbenzene	7.5×10^9		Sehested and Holeman (1979)
Nitrobenzene	3.9×10^9		Buxton et al. (1988)
Aniline	1.4×10^{10}		Buxton et al. (1988)
Benzaldehyde	4.4×10^9		Buxton et al. (1988)
Benzoic acid	4.3×10^9		Buxton et al. (1988)
Benzoate ion	5.9×10^9		Buxton et al. (1988)
Benzophenone	8.8×10^9		Buxton et al. (1988)
1,4-Benzoquinone	1.2×10^9		Buxton et al. (1988)
Chlorobenzene	5.5×10^9		Buxton et al. (1988)
Phenol	6.6×10^9		Buxton et al. (1988)
2-methylisoborneol	5.1×10^9		Rosenfeldt et al. (2005)
Geosmin	7.8×10^9		Peter and von Gunten (2007)

Table B-2 Quantum Yields

Compound	Quantum Yield	Wavelength (nm)	Notes	Reference
H_2O	0.42	172		Heit et al. (1998)
H_2O_2	1.1	254		Goldstein et al. (2007)
HOCl	1.0	254		Feng et al. (2007)
OCl^-	1.0	254		Feng et al. (2007)
NH_2Cl	3.46	222		Li and Blatchley (2009)
NH_2Cl	0.62	254		Li and Blatchley (2009)
NH_2Cl	0.21	282		Li and Blatchley (2009)
$NHCl_2$	2.26	222		Li and Blatchley (2009)
$NHCl_2$	1.80	254		Li and Blatchley (2009)
$NHCl_2$	2.25	282		Li and Blatchley (2009)
NCl_3	0.58	222		Li and Blatchley (2009)
NCl_3	1.85	254		Li and Blatchley (2009)
NCl_3	9.50	282		Li and Blatchley (2009)
$(CH_3)_2NNO$	0.32 (pH 7)	254	q.y. pH dependent	Lee et al. (2005)
$Fe^{III}(OH)^{2+}$	0.24	300		Nadtochenko and Kiwi (1998)
$Fe^{III}(C_2O_4)^{3-}$	1.39	254	actinometer	Goldstein and Rabani (2008); Bolton et al. (2011)
I^-/IO_3^-	0.69 (23.5°C)	254	actinometer – q.y. temperature dependent	Bolton et al. (2011); Goldstein and Rabani (2008)
Atrazine	0.033	254		Bolton and Stefan (2002)
Uridine	0.020	200–300	actinometer	Jin et al. (2006)

REFERENCES

Ashton, L., G.V. Buxton, and C.R. Stuart. 1995. Temperature Dependence of the Rate of Reaction of OH with Some Aromatic Compounds in Aqueous Solution. Evidence for the Formation of a π-complex Intermediate? *Jour. Chem. Soc. Faraday Trans.* 91:1631–1633.

Bolton, J.R., M.I. Stefan, P.-S. Shaw, and K.R. Lykke. 2011. Determination of the Quantum Yields of the Potassium Ferrioxalate and Potassium Iodide–Iodate Actinometers and a Method for the Calibration of Radiometer Detectors, *Jour. Photochem. Photobiol. A: Chem.* 222:166–169.

Bolton, J.R. and M.I. Stefan. 2002. Fundamental Photochemical Approach to the Concepts of Fluence (UV Dose) and Electrical Energy Efficiency in Photochemical Degradation Reactions. *Res. Chem. Intermed.* 28(7-9):857–870.

Buxton, G.V., C.L. Greenstock, W.P. Helman, and A.B. Ross. 1988. Critical Review of Rate Constants for Reactions of Hydrated Electrons, Hydrogen Atoms and Hydroxyl Radicals (·OH/·O$^-$) in Aqueous Solution. *Jour. Phys. Chem. Ref. Data* 17:513–886.

Chin, M. and P.H. Wine. 1994. A Temperature-dependent Competitive Kinetics Study of the Aqueous-phase Reactions of OH Radicals with Formate, Formic Acid, Acetate, Acetic Acid and Hydrated Formaldehyde. In *Aquatic and Surface Photochemistry*, G.R. Helz, R.G. Zepp, and D.G. Crosby, Eds. Boca Raton, FL: Lewis Publishers.

Christensen, H., K. Sehested, and H. Corfitzen. 1982. Reactions of Hydroxyl Radicals with Hydrogen Peroxide at Ambient and Elevated Temperatures. *Jour. Phys. Chem.* 86:1588–1590.

Eibenberger, J. 1980. Pulse Radiolytic Investigations Concerning the Formation and the Oxidation of Organic Radicals in Aqueous Solutions. Ph.D. Thesis, Vienna Univ., Vienna, Austria.

Feng, Y., D.W. Smith, and J.R. Bolton. Photolysis of Aqueous Free Chlorine Species (HOCl and OCl–) with 254 nm Ultraviolet Light. *Jour. Environ. Eng. Sci.* 6:277–284.

Getoff, N., F. Schwoerer, V.M. Markovic, K. Sehested, and S.O. Nielsen. 1971. Pulse Radiolysis of Oxalic acid and Oxalates. *Jour. Phys. Chem.* 75:749–755.

Goldstein, S. and J. Rabani. 2008. The Ferrioxalate and Iodide–Iodate Actinometers in the UV Region. *Jour. Photochem. Photobiol. A: Chem.* 193:50–55.

Goldstein, S., D. Aschengrau, Y. Diamant, and J. Rabani. 2007. Photolysis of Aqueous H2O2: Quantum Yield and Applications for Polychromatic UV Actinometry in Photoreactors. *Environ. Sci. Technol.* 41:7486–7490.

Grigor'ev, A.E., I.E. Makarov, and A.K. Pikaev. 1987. Formation of Cl2– in the Bulk Solution During the Radiolysis of Concentrated Aqueous Solutions of Chlorides. *High Energy Chem.* 21:99–102.

Haag, W.R. and C.C.D. Yao. 1992. Rate Constants for Reaction of Hydroxyl Radicals with Several Drinking Water Contaminants. *Environ. Sci. Technol.* 26:1005–1013.

Heit, G, A. Neuner, P.-Y. Saugy, and A.M. Braun. 1998. Vacuum-UV (172 nm) Actinometry: The Quantum Yield of the Photolysis of Water. *Jour. Phys. Chem. A*, 102:5551–5561.

Jiang, P.-Y., Y. Katsumura, R. Nagaishi, M. Domae, K. Ishikawa, K. Ishigure, and Y. Yoshida. 1992. Pulse Radiolysis Study of Concentrated Sulfuric Acid Solutions. Formation Mechanism, Yield and Reactivity of Sulfate Radicals. *Jour. Chem. Soc. Faraday Trans.* 88:1653–1658.

Jin, S., A.A. Mofidi, and K.G. Linden. 2006. Polychromatic UV Fluence Measurement Using Chemical Actinometry, Biodosimetry, and Mathematical Techniques. *Jour. Environ. Eng.* 132:831–841.

Lee, C., W. Choi, and J. Yoon, 2005. UV Photolytic Mechanism of N-Nitrosodimethylamine in Water: Roles of Dissolved Oxygen and Solution pH, *Environ. Sci. Technol.* 39:9702–9709.

Li, J. and F.W. Blatchley III. 2009. UV Photodegradation of Inorganic Chloramines. *Environ. Sci. Technol.* 43:60–65.

Maruthamuthu, P. and P. Neta. 1978. Phosphate Radicals. Spectra, Acid-base Equilibria, and Reactions with Inorganic Compounds. *Jour. Phys. Chem.* 82:710–713.

Motohashi, N. and Y. Saito. 1993. Competitive Measurement of Rate Constants for Hydroxyl Radical Reactions Using Radiolytic Hydroxylation of Benzoate. *Chem. Pharm. Bull.* 41:1842–1845.

Nadtochenko, V.A. and J. Kiwi. 1998. Photolysis of FeOH^{2+} and FeCl^{2+} in Aqueous Solution. Photodissociation Kinetics and Quantum Yields. *Inorg. Chem.* 37:5233–5238.

Peter, A. and U. von Gunten. 2007. Oxidation Kinetics of Selected Taste and Odor Compounds During Ozonation of Drinking Water. *Environ. Sci. Technol.* 41:625–631.

Roder, M., L. Wojnarovits, and G. Foldiak. 1990. Pulse Radiolysis of Aqueous Solutions of Aromatic Hydrocarbons in the Presence of Oxygen. *Radiat. Phys. Chem.* 36:175–176.

Rosenfeldt, E.J., B. Melcher, and K.G. Linden. 2005. UV and UV/H$_2$O$_2$ Treatment of Methylisoborneol (MIB) and Geosmin in Water. *Jour. Water Supply: Res. Technol. – AQUA* 54(7):423–434.

Schuchmann, M.N. and C. von Sonntag. 1988. The Rapid Hydration of the Acetyl Radical. A Pulse Radiolysis Study of Acetaldehyde in Aqueous Solution. *Jour. Am. Chem. Soc.* 110:5698–5701.

Sehested, K. and J. Holcman. 1979. Radical Cations of Ethyl-, Isopropyl- and Tert-Butylbenzene in Aqueous Solution. *Nuklenika* 24:941–950.

Sehested, K., H. Corfitzen, H.C. Christensen, and E.J. Hart. 1975. Rates of Reaction of O–, OH, and H with Methylated Benzenes in Aqueous Solution. Optical Spectra of Radicals. *Jour. Phys. Chem.* 79:310–315.

Watts, M.J. and K.G. Linden. 2007. Chlorine Photolysis and Subsequent OH Radical Production During UV Treatment of Chlorinated Water. *Water Res.* 41:2871–2878.

Wink, D.A., R.W. Nims, M.F. Desrosiers, P.C. Ford, and L.K. Keefer. 1991. A Kinetic Investigation of Intermediates Formed During the Fenton Reagent Mediated Degradation of N-nitrosodimethylamine: Evidence for an Oxidative Pathway not Involving Hydroxyl Radical. *Chem. Res. Toxicol.* 4:510–512.

Zehavi, D. and J. Rabani. 1972. The Oxidation of Aqueous Bromide Ions by Hydroxyl Radicals. A Pulse Radiolytic Investigation. *Jour. Phys. Chem.* 76:312–319.

Appendix C

Calculation of Fraction of UV Absorbed

For the UV/H_2O_2 AOT, it is important to be able to calculate the fraction of UV absorbed by H_2O_2. The method proposed here can also be used for other AOTs, such as the UV/chlorine AOT, by inserting the appropriate molar absorption coefficients. The method is illustrated by example in a format that can easily be set up as an Excel spreadsheet. In the example, the concentration of H_2O_2 is 25 mg/L = 0.735 mM and the effective path length is 14.0 cm. Table C-1 illustrates the calculations. Column A shows the wavelengths of each band; column B lists the H_2O_2 molar absorption coefficients; column C lists the absorption coefficients (cm^{-1}) of the water without the presence of the H_2O_2; column D is the total absorbance arising from H_2O_2 (value in column B × 0.000735 × 14); column E is the total absorbance arising from the water (value in column C × 14); column F is the relative photon flux from the medium pressure lamp (values sum to 1.000); column G is the total absorbed relative photon flux [value in column F × (1 − 10^{-A}), where $A = A(\text{water}) + A(H_2O_2)$]; and column H is the relative photon flux absorbed by H_2O_2 {value in column G × $A(H_2O_2)/[A(H_2O_2) + A(\text{water})]$}. In this example, the fraction of the photon flux absorbed by H_2O_2 is 0.1174.

Table C-1 Calculations for the fraction of UV absorbed by H_2O_2

Wavelength Band (nm)	H_2O_2 Molar Adsorption Coefficient (M^{-1} cm^{-1})	a(water) (cm^{-1})	Total A Arising from H_2O_2	Total A Arising from Water	Relative Lamp Photon Flux	Relative Photon Flux Absorbed	Relative Photon Flux Absorbed by H_2O_2
200	190.34	1.544	1.96	21.61	0.0163	0.0163	0.0014
201	184.76	1.532	1.90	21.44	0.0150	0.0150	0.0012
202	180.68	1.520	1.86	21.28	0.0216	0.0216	0.0017
203	175.52	1.508	1.81	21.12	0.0174	0.0174	0.0014
204	171.01	1.496	1.76	20.95	0.0168	0.0168	0.0013
205	166.07	1.484	1.71	20.78	0.0150	0.0150	0.0011
206	161.77	1.472	1.67	20.61	0.0131	0.0131	0.0010
207	156.83	1.460	1.61	20.44	0.0117	0.0117	0.0009

Table continues next page

Table C-1 Calculations for the fraction of UV absorbed by H_2O_2 (continued)

Wavelength Band (nm)	H_2O_2 Molar Adsorption Coefficient (M^{-1} cm^{-1})	a(water) (cm^{-1})	Total A Arising from H_2O_2	Total A Arising from Water	Relative Lamp Photon Flux	Relative Photon Flux Absorbed	Relative Photon Flux Absorbed by H_2O_2
208	151.67	1.447	1.56	20.26	0.0116	0.0116	0.0008
209	147.81	1.432	1.52	20.04	0.0108	0.0108	0.0008
210	143.51	1.412	1.48	19.77	0.0113	0.0113	0.0008
211	137.71	1.385	1.42	19.39	0.0105	0.0105	0.0007
212	133.20	1.350	1.37	18.90	0.0100	0.0100	0.0007
213	127.83	1.304	1.32	18.26	0.0092	0.0092	0.0006
214	123.96	1.249	1.28	17.49	0.0091	0.0091	0.0006
215	119.45	1.186	1.23	16.60	0.0091	0.0091	0.0006
216	115.37	1.116	1.19	15.62	0.0086	0.0086	0.0006
217	111.28	1.040	1.15	14.56	0.0095	0.0095	0.0007
218	106.77	0.960	1.10	13.45	0.0092	0.0092	0.0007
219	102.05	0.879	1.05	12.31	0.0096	0.0096	0.0008
220	98.61	0.799	1.02	11.18	0.0099	0.0099	0.0008
221	94.74	0.722	0.98	10.10	0.0102	0.0102	0.0009
222	90.66	0.650	0.93	9.10	0.0102	0.0102	0.0010
223	87.01	0.586	0.90	8.20	0.0111	0.0111	0.0011
224	83.14	0.528	0.86	7.39	0.0108	0.0108	0.0011
225	80.13	0.474	0.82	6.64	0.0114	0.0114	0.0013
226	76.27	0.424	0.79	5.94	0.0113	0.0113	0.0013
227	72.83	0.375	0.75	5.25	0.0107	0.0107	0.0013
228	69.61	0.326	0.72	4.57	0.0107	0.0107	0.0015
229	65.65	0.279	0.68	3.91	0.0105	0.0105	0.0015
230	62.33	0.235	0.64	3.30	0.0113	0.0113	0.0018
231	59.53	0.197	0.61	2.75	0.0096	0.0096	0.0017
232	56.34	0.165	0.58	2.31	0.0095	0.0095	0.0019
233	53.60	0.141	0.55	1.98	0.0091	0.0090	0.0020
234	50.93	0.124	0.52	1.74	0.0094	0.0093	0.0022
235	48.33	0.112	0.50	1.57	0.0096	0.0095	0.0023
236	46.09	0.103	0.47	1.44	0.0064	0.0064	0.0016
237	43.45	0.095	0.45	1.33	0.0067	0.0066	0.0017
238	41.11	0.086	0.42	1.21	0.0082	0.0080	0.0021
239	39.64	0.077	0.41	1.08	0.0060	0.0058	0.0016
240	37.66	0.069	0.39	0.96	0.0082	0.0078	0.0022
241	35.55	0.061	0.37	0.86	0.0046	0.0043	0.0013
242	33.89	0.055	0.35	0.77	0.0040	0.0037	0.0011
243	32.10	0.051	0.33	0.71	0.0039	0.0035	0.0011

Table continues next page

Table C-1 Calculations for the fraction of UV absorbed by H$_2$O$_2$ (continued)

Wavelength Band (nm)	H$_2$O$_2$ Molar Adsorption Coefficient (M^{-1} cm^{-1})	a(water) (cm^{-1})	Total A Arising from H$_2$O$_2$	Total A Arising from Water	Relative Lamp Photon Flux	Relative Photon Flux Absorbed	Relative Photon Flux Absorbed by H$_2$O$_2$
244	30.24	0.048	0.31	0.67	0.0040	0.0036	0.0011
245	28.71	0.047	0.30	0.65	0.0044	0.0039	0.0012
246	26.92	0.046	0.28	0.64	0.0052	0.0046	0.0014
247	25.58	0.045	0.26	0.63	0.0076	0.0067	0.0020
248	24.17	0.044	0.25	0.61	0.0171	0.0147	0.0042
249	22.76	0.042	0.23	0.59	0.0076	0.0065	0.0018
250	21.61	0.041	0.22	0.57	0.0042	0.0035	0.0010
251	20.40	0.039	0.21	0.55	0.0047	0.0039	0.0011
252	19.25	0.038	0.20	0.53	0.0078	0.0063	0.0017
253	18.09	0.036	0.19	0.51	0.0132	0.0105	0.0028
254	17.14	0.035	0.18	0.49	0.0053	0.0041	0.0011
255	16.11	0.034	0.17	0.48	0.0076	0.0059	0.0015
256	15.22	0.033	0.16	0.47	0.0147	0.0112	0.0028
257	14.32	0.032	0.15	0.45	0.0199	0.0149	0.0037
258	13.36	0.032	0.14	0.44	0.0209	0.0154	0.0037
259	12.66	0.031	0.13	0.43	0.0184	0.0134	0.0031
260	11.89	0.030	0.12	0.42	0.0160	0.0114	0.0026
261	11.19	0.029	0.12	0.41	0.0130	0.0091	0.0020
262	10.61	0.029	0.11	0.40	0.0107	0.0074	0.0016
263	10.10	0.028	0.10	0.40	0.0103	0.0071	0.0015
264	9.27	0.028	0.10	0.39	0.0136	0.0091	0.0018
265	8.70	0.027	0.09	0.38	0.0300	0.0199	0.0038
266	8.31	0.027	0.09	0.37	0.0164	0.0107	0.0020
267	7.80	0.026	0.08	0.37	0.0074	0.0048	0.0009
268	7.35	0.026	0.08	0.36	0.0068	0.0043	0.0007
269	6.84	0.025	0.07	0.35	0.0089	0.0055	0.0009
270	6.33	0.025	0.07	0.35	0.0118	0.0072	0.0011
271	6.01	0.024	0.06	0.34	0.0067	0.0040	0.0006
272	5.69	0.024	0.06	0.33	0.0055	0.0033	0.0005
273	5.31	0.023	0.05	0.33	0.0052	0.0030	0.0004
274	4.92	0.023	0.05	0.32	0.0054	0.0031	0.0004
275	4.67	0.022	0.05	0.31	0.0092	0.0052	0.0007
276	4.28	0.022	0.04	0.30	0.0071	0.0039	0.0005
277	4.09	0.021	0.04	0.30	0.0050	0.0027	0.0003
278	3.77	0.021	0.04	0.29	0.0050	0.0026	0.0003
279	3.52	0.020	0.04	0.29	0.0067	0.0035	0.0004

Table continues next page

Table C-1 Calculations for the fraction of UV absorbed by H_2O_2 (continued)

Wavelength Band (nm)	H_2O_2 Molar Adsorption Coefficient (M^{-1} cm^{-1})	a(water) (cm^{-1})	Total A Arising from H_2O_2	Total A Arising from Water	Relative Lamp Photon Flux	Relative Photon Flux Absorbed	Relative Photon Flux Absorbed by H_2O_2
280	3.26	0.020	0.03	0.28	0.0199	0.0102	0.0011
281	3.01	0.020	0.03	0.28	0.0135	0.0068	0.0007
282	2.88	0.019	0.03	0.27	0.0058	0.0029	0.0003
283	2.69	0.019	0.03	0.27	0.0045	0.0022	0.0002
284	2.49	0.019	0.03	0.26	0.0043	0.0021	0.0002
285	2.30	0.018	0.02	0.26	0.0046	0.0022	0.0002
286	2.05	0.018	0.02	0.25	0.0045	0.0021	0.0002
287	1.92	0.018	0.02	0.25	0.0043	0.0020	0.0001
288	1.85	0.018	0.02	0.25	0.0046	0.0021	0.0002
289	1.73	0.017	0.02	0.24	0.0125	0.0056	0.0004
290	1.66	0.017	0.02	0.24	0.0085	0.0038	0.0003
291	1.47	0.017	0.02	0.23	0.0044	0.0019	0.0001
292	1.41	0.016	0.01	0.23	0.0068	0.0029	0.0002
293	1.15	0.016	0.01	0.23	0.0059	0.0025	0.0001
294	1.09	0.016	0.01	0.23	0.0041	0.0017	0.0001
295	1.02	0.016	0.01	0.22	0.0044	0.0018	0.0001
296	0.96	0.016	0.01	0.22	0.0154	0.0063	0.0003
297	0.90	0.015	0.01	0.22	0.0295	0.0120	0.0005
298	0.83	0.015	0.01	0.21	0.0097	0.0039	0.0002
299	0.77	0.015	0.01	0.21	0.0054	0.0022	0.0001
300	0.64	0.015	0.01	0.21	0.0052	0.0020	0.0001
				Totals	1.0000	0.7933	0.1174

Index

NOTE: *f.* indicates figure, *t.* indicates table, *n.* indicates note

A

absorption of light, 14–19
  See also UV transzmittance (UVT)
advanced oxidation technologies (AOTs)
  advantages of, 5–6
  applications for, 2–3
  collimated beam experiments, 25–26, 27*f.*–28*f.*
  comparison of efficiencies, 38, 38*f.*
  comparison of types, 58*t.*
  dark homogeneous, 55–56
  defined, 1
  degradation mechanisms, 24–25
  direct photolysis, 51–55, 52*f.*
  disadvantages of, 6
  figures-of-merit, 34–39
  government regulations, 4–5, 5*t.*
  history of, 3*t.*
  homogeneous advanced reduction processes, 57
  light-driven heterogeneous, 56–57
  light-driven homogeneous, 51
  merry-go-round reactor experiments, 33
  O_3/H_2O_2 treatment of MTBE, 45–46, 45*f.*–48*f.*, 48
  photodegradation of NDMA, 39–40, 39*f.*
  photo-Fenton's process, 44–45, 44*f.*–45*f.*
  by-products, 79–82
  stirred tank experiments, 33
  UV/H_2O_2 treatment of MTBE, 40–41, 41*f.*–43*f.*
  See also design considerations; equipment
aging. *See* fouling-aging factor
air flow, 104
algal blooms, 2, 125–126
Anglian Water, United Kingdom, 127, 129, 129*f.*
AOTs. *See* advanced oxidation technologies (AOTs)
assimilable organic carbon (AOC), 81

B

band absorption, 17–18, 18*f.*
Beer-Lambert Law, 15–16, 26
bench-scale testing, 109
biodegradable organic carbon (BDOC), 81
biologically active carbon (BAC), 101
black-body emission, 64*n.*
bromate, 80–81
by-products
  precursors, 93
  regulated, 4, 79–81
  unregulated, 81–82

C

case studies
  micropollutants, 126–132
  reuse water, 132–135
  taste-and-odor compounds, 125–128
chemical feed systems, 99–101, 118–119, 121–122
chloramines, 78
chlorate, 82

chlorine, 77–78, 101, 144
chloropicrin, 81
collimated beam experiments, 25–26, 27f.–28f.
colors, of light, 11–12
cooling systems, 98
cost analysis, 108

D

dark homogeneous AOTs, 55–56
degradation mechanisms, 24–25
design considerations
   chemical feed systems, 99–101
   chemical management, 122–124
   costs, 108
   electric power systems, 104–106
   energy management, 122–123
   engineering consulting firms, 147
   filtered water systems, 87t.
   flow rate, 92–93
   fouling-aging factor, 96–97
   groundwater systems, 87–88
   hydraulics, 103–104
   hydrogen peroxide quenching, 101–103
   hydroxyl radical scavenging demand, 93
   information needs, 147–148
   ozone systems, 97–99, 107
   by-product precursors, 93
   redundancy, 94
   reuse applications, 88–89
   target contaminants, 91–92
   treatability testing, 108–110
   treatment goals, 91
   turndown capacity, 93–94
   unfiltered water systems, 85–86, 86–87
   UV system site layout, 106–107
   UVT, 95
direct photolysis, 51, 52f.
disinfectant residual, 77–78
disinfection applications, 4
disinfection by-products (DBPs), 79
disinfection credit, 86
dose per log (D_L) concept, 38–39

E

electrical energy dose (EED), 34, 38f.
electrical energy per mass (E_{EM}), 34–36
electrical energy per order (E_{EO}), 34–38, 35f., 36t., 37f.
electrical power, 104–106, 137–138
emission of light, 14
endocrine-disrupting compounds, 2
engineering consulting firms, 147–148
equipment
   closed-pipe systems, 62, 63f.
   content for specifications, 149t.–151t.
   design considerations, 61
   electric power systems, 104–106
   hydrogen peroxide quenching system, 71–72
   liquid oxygen (LOX) storage, 70, 145
   manufacturer questions, 149
   nitrogen boost, 70
   open-channel systems, 61
   oxidant chemical systems, 71
   ozone contactors, 69
   ozone destruct units, 69–70
   ozone generators, 68–69, 97–98
   ozone instrumentation, 71t.
   ozone monitoring, 70–71
   pump maintenance, 119
   sleeves, 67, 78–79, 96–97
   UV lamps, 62–66, 63t., 65f., 65t.–66t., 96–97, 138–141, 140t.
   UV monitoring, 68
   UV sensors, 66–67, 96–97, 120
   vaporizers, 70
examples of AOTs
   O_3/H_2O_2 treatment of MTBE, 45–46, 45f.–48f., 48
   photodegradation of NDMA, 39–40, 39f.

photo-Fenton's process, 44–45, 44*f.*–45*f.*
UV/H$_2$O$_2$ treatment of MTBE, 40–41, 41*f.*–43*f.*
excited-state processes, 17–18

F

facilities
   filtered water systems, 85–86, 87*t.*
   groundwater systems, 87–88
   regulatory issues, 151
   site layout considerations, 106–107
   start-up steps, 111–115
   unfiltered water systems, 86–87
   water reuse, 88–89
   See also design considerations
flow rate, 92–93
fluence-based rate constants, 32–33
fouling-aging factor, 96–97

G

germicidal range, 13
granular activated carbon (GAC), 71–72, 85, 101–103, 103*t.*
groundwater systems, 87–88
Guidelines for Water Reuse, 5

H

hazardous waste disposal, 141–143
head loss, 103–104
health and safety. *See* safety issues
hydraulic grade line, 86
hydrogen peroxide (H$_2$O$_2$)
   and disinfectant residual, 77–78
   off-gassing, 100, 119
   quenching, 71–72, 101–103, 102*f.*, 103*t.*
   safety issues, 144
hydroxyl radicals
   AOT mechanisms, 24–25, 33–34
   properties of, 23–24

scavenging demand, 76–77, 93
See also photolysis

I

injection systems, 118–119, 121–122

K

K-Water, South Korea, 125–126, 127*f.*–128*f.*, 127*t.*–128*t.*

L

lamps
   aging, 96
   break causes and prevention methods, 139, 140*t.*
   classification of UV, 63*t.*
   comparison of, 65–66, 66*t.*
   disposal of, 139, 141
   effects on AOT system efficiency, 74–75
   excilamps, 65, 65*t.*
   flash, 64–65
   gas discharge, 62
   mercury vapor, 62–64, 65*f.*
   off-line breaks, 139, 142
   online breaks, 139, 140*t.*, 142–143
   power supplies for, 67–68
   safety issues, 138–141
   sleeve fouling, 78–79, 96–97
light, defined, 11*n.*
light-driven heterogeneous AOTs, 56–57
light-driven homogeneous AOTs, 51
liquid oxygen (LOX), 70, 145

M

management considerations, 147–151
maximum contaminant levels (MCLs), 4
mercury
   regulatory requirements, 139–141, 141*t.*
   release response, 141–143, 144*t.*
merry-go-round reactor experiments, 33

micropollutants
   treatment case studies, 126–132
   treatment of, 2
molecular absorption, 17–18, 18f.
monitoring
   frequencies, 121, 122t., 123t.
   mercury release, 143, 144t.
   ozone systems, 121, 123t.
   UV systems, 120–121, 122t.
monochromatic light sources, 29–33
MTBE (methyl-t-butyl ether)
   O_3/H_2O_2 treatment example, 45–46, 45f.–48f., 48
   UV/H_2O_2 treatment example, 40–41, 41f.–43f.

N

natural organic matter (NOM), 76
NDMA (N-nitrosodimethylamine
   photodegradation example, 39–40, 39f.
   reuse treatment case study, 132–135
near infrared range, 12
nitrite, 80

O

O&M manual, 112
Occupational Safety and Health Administration (OSHA), 137, 141
off-gassing, 100, 119, 144–145
OH radicals. See hydroxyl radicals
1,4-dioxane
   state guidelines, 5t.
   treatment case study, 129–133
operations & maintenance (O&M)
   manual, 112
   oxidant injection systems, 118–119
   ozone systems, 117–118
   task examples, 116t.–117t.
   UV systems, 115
oxidants
   dosing system equipment, 71

free chlorine, 77–78
hydroxyl radicals, 23–25
injection system O&M tasks, 118–119, 121–122
oxidation
   wet air process, 56
   See also advanced oxidation technologies (AOTs)
oxygen feed systems, 97
ozone systems
   AOT design criteria, 97–99
   diffusion system, 98–99
   and disinfectant residual, 78
   equipment, 68–71
   functional testing, 113–114
   generator alternatives, 97–98
   injection advantages and disadvantages, 99t.
   monitoring, 121, 123t.
   O&M tasks, 117–118
   safety issues, 144–145, 145t.
   site layout considerations, 107
   target dose, 97, 98f.

P

peroxone (O_3/H_2O_2), 55, 127, 129
pesticide treatment case study, 127, 129
pH, effects on AOT process, 77
photochemistry
   laws of, 19–20
   polychromatic processes, 33
   reference material, 11
   sensitized reactions, 28–29, 33–34
   spectral ranges, 12–14
   See also photolysis
photo-Fenton's process, 44–45, 44f.–45f., 54–56
photolysis
   direct, 51–55, 52f.
   reactions, 29–33, 32f.
   of sulfur-containing ions, 57

photons, 12
pilot-scale testing, 109
Planck Law, 12
polychromatic light sources, 33
power, electric, 104–106
processes
   dark Fenton's, 55–56
   excited-state and absorption, 17–18
   homogeneous advanced reduction, 57
   O_3/H_2O_2 (peroxone), 55, 127, 129
   photolysis of sulfur-containing ions, 57
   radiation, 56
   sonolysis, 56
   UV/chlorine, 53
   UV/H_2O_2, 53, 129–132, 130*f.*, 131*t.*, 132*f.*–133*f.*
   UV/iodide, 57
   UV/O_3, 52
   UV-vis/Fenton's, 54–55
   VUV water photolysis, 52–53
   wet air oxidation, 56
pumps, maintenance, 119

Q

quality
   power, 105
   water, 73–82, 93
quantum yield, 18–19
quenching, 71–72, 101–103, 121–122

R

radiation, 56
rate expressions, 30–31
reclamation. *See* reuse
recycling (wastewater). *See* reuse
redundancy, 94
reflection, 14, 15*f.*
refraction, 15, 15*f.*
refractive index, 14
regulatory issues
   disinfection credit, 86

mercury, 139–141, 141*t.*
1,4-dioxane state guidelines, 5*t.*
performance testing, 115
permitting, 151
water reuse, 4–5, 6*t.*
Resource Conservation and Recovery Act (RCRA), 139, 141
reuse
   AOT applications, 3, 88–89
   regulatory issues, 4–5, 6*t.*
   treatment case study, 132–134, 133*t.*, 134*f.*–135*f.*

S

Safe Drinking Water Act (SDWA), 4, 139
safety issues
   burns, 138
   chemicals, 143–145
   electrical, 137–138
   mercury, 139–143
   UV light exposure, 138
scattering of light, 16
scavenging demand, 76–77, 93
sensitized photochemical reactions, 28–29, 33–34
sensors, 66–67, 96–97
sleeves, 67, 78–79, 96–97
sodium hypochlorite, 100–103, 102*f.*, 103*t.*
solar blind sensors, 66
sonolysis, 56
spectral ranges, 12–14, 13*f.*, 13*t.*
spectrophotometers, 16–17
staffing, training, 114
steady-state approximation, 28–29
stirred tank experiments, 33
storage tanks, maintenance, 118–119
surface water systems
   filtered applications, 85–86, 87*t.*
   unfiltered applications, 86–87

T

target contaminants, 91–92
taste-and-odor compounds
   treatment case study, 125–126, 127*f.*–128*f.*, 127*t.*–128*t.*
   treatment of, 2
testing
   functional, 112–114
   performance, 112, 114–115, 123
   treatability, 108–110
total organic carbon (TOC), 75
toxicity, 82
training, 114
transmission of light, 14–17
   See also UV transmittance (UVT)
treatability testing, 108–110
trichloropropanone (TCP), 81–82
Tucson International Airport Area Groundwater Remediation Project (TARP), Arizona, 129–132, 130*f.*, 131*t.*, 132*f.*–133*f.*
turbidity, effects on AOT process, 77
turndown capacity, 93–94

U

ultraviolet (UV) light
   defined, 11
   direct photolysis processes, 52
   exposure to, 138
   subranges, 13
US Environmental Protection Agency (USEPA), 4–5, 86, 111
utilities. *See* facilities
UV Disinfection Guidance Manual, 96
UV light. *See* ultraviolet (UV) light
UV systems
   AOT design criteria, 94–97
   equipment, 61–68
   full-scale reactors, 37*f.*, 38
   functional testing, 113
   O&M tasks, 115
   sensor calibration, 120
   site layout considerations, 106–107
UV transmittance (UVT)
   analyzer calibration, 121
   defined, 16
   design considerations, 95
   and water quality, 73–75, 75*t.*

V

vacuum UV (VUV), 13–14
   water photolysis process, 52–53
vaporizers, 70
visible light, 11
visible range, 12
volatile organic compounds (VOCs), 2

W

wastewater. *See* reuse
Water Campus, Scottsdale, Arizona, 132–135, 133*t.*, 134*f.*–135*f.*
water hammer, 86
wavelength ranges, 12